Global Strategies of Clean Environment Safe Earth Disaster Management Sustainable Development and Quality Life

(National and International Obligations and Priorities)

N.P. RAO

ATLANTIC PUBLISHERS AND DISTRIBUTORS

Published by
ATLANTIC PUBLISHERS AND DISTRIBUTORS
B-2, Vishal Enclave, New Delhi-110 027
Phones 5413460, 5429987

Sales Office
4346/4C, Ansari Road, Darya Ganj,
New Delhi-110 002
Phones 3273880, 3285873, 3280451
Fax 91-11-3285873
E-mail : gupta.atlantc@axcess.net.in

ISBN 81-7156-740-1

Typeset at
APD Computer Graphics, Delhi

Printed in India at
Mehra Offset Printers, Delhi

Preface

Planners, thinkers, environmentalists, educationists, scientists and a host of others all over the world are concerned about how to save the Earth and environment. Arresting environmental depredation and steps to correct the mistakes of the past call for immediate attention.

In the vast arena of the world, with various ideologies, shapes, sizes, colours and stages of development, it is difficult to tackle the issues of creating a sustainable world, based on equity and justice. We should involve people at the grassroot level to visualise a world that would be livable and enjoyable for all its 5.4 billion people.

The urgency lies in the fact that by the turn of the century, there would be 6.2 billion people in the world, to claim their share of fast depleting natural resources, food, water and even ozone. Seventy-three per cent of these people live in the developing countries in Asia, Africa and South America, mostly in sub-human conditions.

The plundered earth is in ruins, with rising crime, slums, denuded forests and devastated land. The people are however keen to rebuild their countries and to usher in a new world order.

Representatives of slums invoke the marginalised urban population from the cities around the world to prepare a plan which "gives them a rightful place in the world beyond." Their interest goes beyond dealing with the critical problems of environment, from climate change to ozone depletion, soil erosion and the litany of human created destructions during the last 300 years in the name of progress and development, to more fundamental issues which are at the root of all earthly problems. These are the problems of development : population, the role of women, poverty, technology transfer, unfavourable trade systems, the Third World debt of $ 1.3 trillions, lifestyles of the Northern hemisphere, moral and ethical issues, piling of armaments (at an annual cost of $ 1.4 trillions) and its impact on environment and development and archaic geographical boundaries.

These are the same that have made this world so sick. We have to conceive a complementary, comprehensive and holistic plan to carry the message of hope and cheer to the entire world.

People of developing countries only seek sheer means of survival — water, food, renewable sources of energy, education and means of livelihood. Their only wish is to live in harmony with the land, trees, birds and animals as they have done since human beings evolved as homosapiens, four millions years ago. They hope to create their own future, instead of leaving it to destiny. It will be created only by those who will provide every child a livable, sustainable and enjoyable future so that evolution progresses without violence and bloodshed.

When it was being contemplated to write this book a devastating earthquake rocked Maharastra and adjoining States in India, killing over 10,000 people, ruining 30,000 families and flattening the affected villages, turning them into graves. Fighting the disaster included removal of the deadbodies, saving those alive with medicines, providing shelter, food and water, and reconstructing dwelling houses to rehabilitate them. Governmental, non-governmental world bodies and Governments and people of other countries have come forward with due aid. A chapter on Management of Natural Disasters including earthquakes is therefore incorporated which is of topical importance and impelled the Author and prompted him to write this book in this International Decade for Natural Disaster Reduction. Similarly it is a frequent phenomenon that monsoons, cyclones, tropical storms, torrential rains hit coastal and lowlying areas killing hundreds of residents, render homeless thousands of people, flood large areas, destroy cattle, submerge crops, cut off communications, disrupt the lines, cut off roads and traffic, inundate the reservoirs, breach dams and tanks, maroon the villages, devastate houses etc. Rehabilitation measures include evacuating the marooned people to safer areas through mechanised boats, supply of food and medicines, releasing grants-in-aid, compensation for losses etc. These and other relief measures are discussed in this book.

A World Bank report released recently says that more than 40,000 people die prematurely every year in India alone because of Air Pollution. Patients with respiratory diseases and allergies have doubled since 1990. Auto exhausts and industrial emissions have dangerously crossed the limits. 21st century will be less in oxygen and more in carbon dioxide and other pollutants. Dust and particles of carbon coated with toxic gases have exceeded the standards of safe levels. Aspirational Bronchopneumonia has become common due to the pollutants from petrol cars and vehicles which throw out/emit hydrocarbons, oxides nitrogen, black smoke and particles. By 2001 carbon dioxide emissions will rise 7 times and hydrocarbons 9 times. Other pollutants will increase 5 times. Lead in petrol affects central nervous system, carbon dioxide reduces the ability of blood to carry oxygen and provokes heart diseases, suspended particles of dust and carbon cause respiratory diseases, Aromatic hydrocarbons from diesel engines cause drowsiness, eyes irritation, cough etc., sulphur dioxide causes bronchial problems, benzene causes lung cancer, Nitrogen oxides cause colds; cough, bronchial irritations etc. The fuel quality is poor and its adulteration is common. Latest technology to improve carbrettors, fuel quality and use of catalytic converters, pure unleaded petrol etc. should go a long way to prevent air pollution from vehicles. Better traffic management, rapid transit systems, better fuels, checks and control and maintenance would help a lot.

The aim is to establish a new world order for the future millennium — an order based on sustainability, equity and justice. For this, the subject of Earth protection and Environment safety is not only of paramount importance but also topical. Mother Earth which provided our life must be saved at any cost. The pages that follow, in this book deal with these issues involved. In the 3 parts of this book, the topics discussed include environmental conservation, protection and preservation, ecological safety, monitoring, checking the damage, waste recycling and re-utilization, hazardous waste disposal, adoption of cleaner technologies for eco-safety, protection from eco-hazards, disasters and climatic and atmospheric changes, natural and man-made disasters and rehabilitation of the affected, global strategies and policies, Earth summit recommendations for implementation, the road ahead and

---

The United Nations Framework convention on climatic change for Developing Countries is concerned with the reduction of Green House Gas (GHG) emissions through International Partnerships and collaborations between developed and developing countries for solving problems of environment, earth safety, and sustainable development. GHG must be reduced particularly in power, transportation, building and agricultural sectors through reduction in fuel consumption, better maintenance of vehicles, cleaner production, low energy substitutes, waste cycling — all intended to reduce the enormous carbon dioxide emissions. Similarly reforestation and appropriate cropping methods could reduce the GHG concentrations in the atmosphere. Likewise, setting up of bio-mass gasifiers and mini-hydel power stations and bio-methanation to produce electricity, besides the use of solar energy for rural electrification are recommended in the Power Sector. These could reduce about 70% of the carbon dioxide emission. As the global climate change affects both the developed and developing countries, there is an immediate need for greater cooperation between the North and South countries to implement the Earth Summit recommendations — particularly reducing the carbon dioxide emissions in industrialised countries which have an impact on the climate. New mechanisms should also be found out to deal with climate change and transfer of technology to developing countries. This should be stepped up through afforestation and bio-diversity.

At the last Rio Earth Summit Conference, new International conventions were adopted on combating climate change, protecting the World's wildlife and plants and resisting the spread of deserts. Goal was also set at Rio to reduce the "Green House Gas" emissions behind global warming to 1990 levels by 2000. Besides, new targets are set in fields ranging from conserving freshwater supplies to limiting chemical pollution. The UN Commission on sustainable Development and the UN General Assembly Special Session — Earth Summit 2 and the National Wildlife Federation are seriously concerned with this matter Cleaning up World's oceans, curbing overfishing and protecting forests are also receiving their attention. The Environment Organisation "Green Peace" has drawn attention of all UN Missions regarding the curbing of overlogging, stemming degradation of forests, need for new International Agreements on Oceans, fresh water and toxic chemicals. The General Assembly is to give a new plan and direction to protect the World's fresh water resources and promote cooperation between countries drawing their supplies from the more than 800 rivers crossing National borders. New controls are going to be adopted on chemical pollution and insuring that such chemicals are *Not* exported without the consent of an importing country, and

forward looking to ensure cleaner and safer surroundings for sustainable development, and better quality of life at local and global level.

Grateful acknowledgements are made to the various publications and the authors who inspired me to write this book. Any suggestions will be incorporated in the future editions.

DR. N.P. RAO

---

restricting the use of 12 dangerous chemicals. At present world's oceans are in acute danger and they are being used as *waste dumps*. Overfishing has also assumed alarming proportions.

What is needed now is that all the world's governments should scrupulously implement the Agreements made and recommendations adopted at the various International, Inter-Governmental, Ministerial, General Assembly and U.N Conferences. These were discussed in the recent Earth Summit.

Contents

PART - III

Introduction

GLOBAL STRATEGIES TO PRESERVE THE EARTH AND ENVIRONMENT OR PERISH—THERE IS NO OTHER OPTION

Since 1990, the anthropogenic pressures on earth multiplied by more than three times, 75 to 80 per cent in the developing world and the balance in the developed. World economy expanded 20 times, consumption of fossil fuels has grown by a factor 30, and industrial production increased by a factor 50. The impact of human activity has led to unsustainability to a level that bioactivity is utilising not the interest of the capital natural resources but living on the resources themselves.

In reality

(i) Man could neither create nor destroy the elements essential as gifts of nature for biomass production. In the endeavour to assemble and manage the elements for production of food, utilities, necessities and luxuries catalysed by technological advances, excessive greed led to mismanagement to make the production system fragile and exhausting the carrying capacity.

(ii) Although the said endeavour is to meet the needs of the growing population yet the polarisation is occular with 30 per cent of the population consuming 70 per cent of resources with high standards of living while the 70 per cent of the population left with the rest of the resources live in abject poverty and hunger. Co-existence without social unrest became impossible in an unequal world.

(iii) A povertycurtain divides the world materially and philosophically. One world is literate, the other illiterate, one industrial and urban, the other largely agrarian and rural, one consumption oriented and the other struggling for survival. We witness maldistribution of the world's resources on a scale where industrialised countries are consuming about 20 times more of the resources per capita than the poor countries. The per capita consumption of developing countries as per cent of the developed is 70 per cent of calories, 55 per cent fat, 6 per cent paper, 8 per cent steel and 7 per cent commercial energy.

The horrors of pollution from both agricultural and industry, the threat of ozone shield depletion, acid rain, extincting of species are being debated all over the world.

(iv) Today the developing world with nascent freedom seeks faster development on the Western model to make up the lost time and ground. The developed world under the threat of climate change, and environmental degradation is struggling to maintain existing standards of living.

Thus the concern of the Third World, India in particular is : Would the achievements in food production and industrial growth continue on a scale to meet the growing demands on sustainable basis? It would be most formidable to address the environmental issues under lack of political commitment, limited financial resources and inadequate training and skills.

Efforts are called for (i) to reduce greenhouse gas emissions, (ii) increase financial and technological flow to the "developing world" for environmental conservation, sustainable development and population control programmes, (iii) reduce the imposed burden of external debts by suitable aid transfer and related mechanisms, (iv) develop indigenous technologies for a blend with the "frontier", provide scientific tools and training to upgrade skills with access to information and expertise — all in all to promote "self-reliance and management".

The relationships between economic developmental goals and natural resources use are fragile. More and more food is to be produced on less and less land area. Energy saturated inputs substituting the land for production is unsustainable. It is on record that 58 ml. ha of irrigated area in India, a third (20 ml. hs) has been lost due to salinisation. Climatic changes are seen as drought years to occur. Biotic impoverishment — the loss of forest area inflicted adverse change especially loss of flora and fauna, soil erosion, silting of dams reducing their capacity for storage and life span. The coastal areas where the livelihood security of 20 per cent of the population resides is under threat of sea level rise for a total wipe out. The population rise is not mere statistics but lamentable realities such as the ratio between the developed and developing world that moved from 1.6 in 1950 to 2.9 by 1990 and would be 3.9 by 2000. Thirty five per cent of the Indian population would be in the highly reproductive age group of 18.35 with limited job opportunities. The per capita land decreased from 0.329 ha of 1950 to 0.173 ha by 1995 with density per sq.km rise to 254 over the same period.

Harmonising economic growth with environmental protection needs a firm acceptance that the two are not antithetical. Economic growth can and must be the engine of envorinmental improvement. It must pay for the technologies of protection and clean up. For the Third World to maintain economic growth for sustainable development in agriculture, industry services and human development, the following must be accepted.

(i) A statutory enactment for land, water and bidiversity conservation and utilisation, not in isolation, but integrated and mutually reinforcing must be put into action.

(ii) Pollution of water, air, land and industrial pollutants must be put under monitoring and control for which finances must be generated from within the country on the principle enacted and operated in the advanced countries, "polluter pays". Agricultural production should not be exempt from this principle. To minimise noise pollution, landing charges for aircraft noise, prorated according to engine type or aircraft weight, should be levied as in Japan and in Sweden.

Subsidies of all kinds must terminate on such utilities vulnerable to waste such as fertiliser, water, etc. with strict adherence to practices that would enhance use efficiency, per unit resource over unit time.

(iii) (a) Protecting the top soil, (b) reforestation, (c) slowing population growth, (d) raising energy use efficiency, (e) development of solar, wind, geothermal and biomass energy, (f) investments for improvements of animal husbandry for energy, meat, milk, wool and manure, these and various other related issues are very pertinent for consideration at global level.

Realising the importance of Disaster prevention, the UN General Assembly has proclaimed the current decade as the "INTERNATIONAL DECADE FOR NATURAL DISASTER REDUCTION (IDNDR) with the objective to reduce loss of life, damage to property and social and economic disruption caused by natural disasters. The quality of water, air, food and habitat on EARTH is dramatically receding, with a simultaneous increase in the temperature of the planet. Natural disasters are causing loss of lives and property, as the resources are depleting rapidly. This vicious circle encircles even the so-called rich, as is evident from increasing aridity, floods, chemical and nuclear hazards and threats, and INTERNATIONAL TERRORISM.

Steps to tackle disasters should therefore be planned well in advance keeping in mind the specific calamities that may strike at a particular area. In most of the cases, *disasters are generated by man's undesirable acts disturbing the natural balance. Due emphasis should accordingly be given for preparations to meet a crisis rather than planning for the post-destruction phase. Similarly cleanliness drives must be launched on a war footing in the coming decade to keep diseases at bay.*

Severity of a calamity can be guaged only by understanding the agony caused to an individual and training for "*acute trauma life support*" (ATLS) should be made necessary for every citizen so that life of a man in danger could be saved.

Steps should include an integrated and swift response to any natural hazard event, loss minimisation through preparedness and preventive measures, reduced vulnerability through concerted action and transfer of technology, and building up community capabilities to recoup from the losses caused by any hazard, providing group insurance for poorer sections of society, and individual insurance to help minimise the need for external assistance in matters of relief and rehabilitation, transfer of new techniques particularly in housing, and ensuring a state of readiness and preparedness in all phases of disaster impact round the year.

GLOBAL CO-OPERATION URGENTLY NEEDED TO AVERT AND COPE UP WITH NATURAL DISASTERS

United Nations Inter-Governmental Oceanographic Commission (IOC) is seized of the need for urgent regional and international cooperation to mitigate the disasters *through exchange of data, transfer of technology and emergency responses. Stepping up of R&D for precise prediction, an integrated approach for better public awareness, a sound communication network to give timely alarm and a quick response management system for evacuation and protective measures are the essential inputs for effective management of natural hazards.* In the case of storm and cyclones, the use of various predictive methods, analysis of the hydrodynamic equations and finding the 'numerical solutions' would be helpful and an important step to operationalize these predictive models is the validation by using the measured data of the severe cyclonic storm. Input parameters include meteorological data, oceanographic parameters, basin characteristics — coastal geometry, surface and bottom stress of water column and hydrological parameters. Alongwith R&D on storm *surge*, steps must be taken to devise suitable mechanism to estimate the hazard potential and its impact on society to formulate the protective measures. A 'storm surge project' is being contemplated exclusively for the Bay of Bengal and the Indian ocean under the joint auspices of the Intergovernmental Oceanographic Commission (IOC), World Meteorological Organisation, (WMO), UNESCO, IHP, UNEP and other allied agencies. A 'Natural Disaster Management Centre' is being proposed in the 9th Five Year Plan. The Hawaii based World Meteorological Organisation tracks the movements of depression to provide pictures through Internet for necessary action against cyclonic storms.

NEW THREAT FROM GLOBAL WARNING WHICH IS ABETTING DISEASES LIKE DENGUE FEVER (PRESENT EPIDEMIC IN DELHI) — IMPACT OF INCREASED TEMPERATURES ON THE EARTH'S CLIMATE

In the latest report issued by the World Health Organisation, a United Nations Agency, the World Meteorological Organisation and the United Nations Environment, millions of people are becoming vulnerable to the adverse effects of global warming resulting in dengue fever, malaria and other communicable diseases. Dengue fever has spread to India, Central and South America besides some countries in South Asia and Africa. (Dr. Jonathan Patl of John Hopkins School of Public Health). Also Malaria has become widespread through the developing world — both the above diseases caused through mosquitoes. Rising temperatures have affected the tiny plankton in the sea that contribute to cause cholera. Heat related deaths have increased in Shanghai, China, Chicago in the US. Increased levels of temperature in sea water would kill aquifers in the low-lying areas particularly in Asia and Africa. Global warming could destabilise the world's climate and its dependent ecosystems posing new and widespread risk to the health and human well-being. Environmental hazards are arising from ecologically disruptive and irreversible global environmental processes. The increased number of heat related deaths in India, midwest United States and Southern Europe and the changes in the geographic range of some vector-borne diseases (due to high temperatures) could not be ignored.

According to Charles Miller of the Environmental Media Services of the US government, the pollution prone human activity is causing a warming of the world and that is why the spring season has been arriving earlier each year in the Northern Hemisphere. There lies therefore the necessity to reduce greenhouse emissions in the developed world to 1990 levels by the year 2000. Steps should be taken to reduce emissions of such greenhouse gases as carbon dioxide, produced by the burning of coal, oil and wood to generate electricity to heat homes, to power cars and trucks and to cook food. An International Team of Scientists have already warned that Earth's temperature could increase by 1-3.5 degrees Celsius, and sea levels by 15-95 centimeters in the next 100 years *unless greenhouse emissions are not reduced.* This has been highlighted in the latest report presented in an International Conference held in Geneva, Switzerland, attended by more than 150 countries of the World to consider measures to reduce the emissions of the greenhouse gases that are preventing some of the heat rising from Earth from dissipating in outer space. Lawrence Kalkstein of the US Environmental Protection Agency and Barry Levy of American Public Health Association have focussed the underlying hazards and urged the governments to take measures to prevent changes in the Earth's climate.

EMISSION OF CARBON DIOXIDE POLLUTANT — A GLOBAL SCENARIO

The World Resources Institute estimates that between 1992 and 1997, global CO_2 emissions increased by 38%. Between 1982 and 1992, there was a 20% increase in total emission of these atmospheric pollutants. In 1992, 22.3 billion tons of CO_2 was released from Industrial activity.

A 1992 United Nations Agreement on Climate Change seeks to control the emission of carbon dioxide, the gas that contributes to global warming.

The US is the largest emitter of CO_2, but two Developing countries — China and India — are also among the 10 largest polluters. However, the World Resources Institute estimates that the OECD countries account for 45% of global emissions and the European Union alone contributes to 13% of the total. Even Kazaahkastan and Saudi Arabia also figure in the list, because of the large amount of gas flared in these major oil producing countries.

A comparison on a *per capita* basis gives better indication of who is responsible for how much CO_3 pollution. The US remains the *biggest* polluter among the listed 25 countries while China is third lowest and India emits the least on *per capita* basis (since the population is high). The *per capita* figures also show that the *developed countries* are the *biggest* polluters (US, Australia, Canada, Germany etc.), while the developing countries account for insignificant amounts of pollution (China, India and Indonesia). Even Brazil, often blamed for destroying vast tracts of forests, emits only a small amount of CO_2. But the East European countries and the oil producing economies remain large polluters even on a *per capita* basis.

As regards Disaster Management relating cyclones, also discussed in this book, the use of early warning systems, setting up more relay stations to relay the bulletins to those living in the cyclone-prone areas, use of sophisticated equipment are recommended by the International Resource Group, World Bank and Environmental Consultants of the World Bank. Use of 'Doppler Radar' and super computers for weather forecast, rescuing fishermen at sea at the time of cyclone warning, utilising the services of the coast guard for rescue work, deploying helicopters for locating the missing boats and men and shifting the latter to safety are the other ways to be adopted. Weather warnings should be framed in such a way that the impending disaster is well-broadcast from time to time. There should be a process of constantly updating our knowledge of warning systems and improving them. Information should be promptly disseminated, and warning bulletins issued timely to avert the disaster, as reiterated by the above UN Agencies.

PART-I

1. Environment and Earth

THE ENVIRONMENT CHALLENGE

A major problem is to solve environment issues by a concrete environment policy.

Environmental issues are a major cause of worry. We should muster all resources to tackle the hazards of environmental pollution caused by industries, automobiles and garbage dumps. Eighty per cent of the electricity is produced using coal with high percentage of sulphur and the emissions have a foul smell, like that of rotten eggs. Many children in the industrialised cities and towns show a high percentage of bronchial problems. In many places drinking water has to be transported from distant areas. We should focus our attention on these basic issues. The levels of sulphur dioxide should be brought down by the use of special filters. The concentration of industries pollutes the rivers.

Automobile exhaust is a major pollutant. While 90 per cent of the automobiles in the West used unleaded petrol, many vehicles in the East still run on leaded gasolene. The unleaded petrol is made cheap to encourage more people to use it, and the leaded petrol is taxed high. Most of the automobiles in Europe are fitted with catalytic converters to minimise the toxic fumes.

Priorities are cleaning, waste management, climate protection and traffic management. Progress has to be made in controlling air-pollution and water protection. We have only one per cent of the total geographic area under strict protection, and we lag behind in the conservation of biological diversity. With a high degree of environmental literacy, we should try to reduce damage to the environmental assets, reduce and reuse wastes and keep the country clean for the generations to come.

The industrial emissions and pesticides problem is also severe, also the use of poor quality brown coal and the heavy dose of chemical fertilizers used pollute the soil. We have to take up some emergency measures for cleaning drinking water.

Integration of environmental aspects in agriculture, transport and energy policies is an important part of this process. The emphasis of the principle will be on recycling packaging material such as bottles, plastic tubes, trays and packaging of specialised products.

Voluntary participation of industries in environmental programmes is a welcome trend. Manufactures should take back all the packaging material and recycle them. This would cut cost to the consumers, and also to the industries besides avoiding environmental pollution. Special programmes for disposing domestic garbage by segregating the waste at house-levels are also recommended.

Increase in ozone level in the atmosphere is another problem. The ozone concentration should be flashed prominently on huge electronic screens at important road junctions in cities for public information. On hottest day of the year, increased levels of ozone, should be shown, and the residents duly warned from exposure. All such protective measures should be enforced.

ENVIRONMENT DAMAGE

The problems should be microised, that is identified in the immediate locality, and tackled systematically. The word 'environment' should itself be dispensed with and substituted with "right to life" because the former had come to be associated with what one expected someone else to do.

Each Indian accounted for the destruction of seven trees annually and planting trees would compensate for it.

It is time to think what kind of legacy should be left behind for posterity. People saved money, jewellery or other articles for the future generations without giving a thought to the kind of environment they were leaving behind.

Meat eating was indirectly harmful to environment. Though each goat may cost a few hundred rupees, its actual value was about Rs. 60,000 because this was the value of the grass it consumed.

Cigarettes, were another source of damage to environment and quality of life because tobacco was grown on rich soils and was a water-intensive crop. Curing of tobacco required wood fuel. The manufacture of cigarette paper used up 100,000 hectares of forests and that of match sticks 15,000 hectares.

It is no good constructing box-like houses which raised room temperature instead of airy ones and using synthetic bags instead of those made of jute. However, going ethnic by wearing cotton and khadi might not always be good. Huge quantities of pesticides were used in growing cotton while dyes used in printing contained harmful chemicals.

For preserving the environment, it had been estimated that if every school going child saved one page every day 40,000 trees could be saved from destruction. Environmental learning should be encouraged. A joint forest management involving the rural poor for protecting forests and improving degraded ones and forest-based industries could be involved in a big way in afforestation programmes.

Plantations should be made compulsory around apartments. Civic laws should be made more stringent. Non-governmental agencies could play a big role in environment protection. Bio-technology and pollution control will go a long way in this direction.

ENVIRONMENT DYNAMICS

Atmospheric and climatic changes imminent on earth, as inevitable consequences of the utterly irresponsible recent anthropogenic activities are posing serious survival concerns for mankind. We should predict global environmental changes and prevent or at least mitigate them. Since three-fourths of the globe are occupied by the developing countries, they should realise their role in this impending doom.

India is responsible for about 4 per cent of the anthropogenic emissions of the greenhouse gases. Although carbon dioxide (CO_2) is known to contribute to 55 per cent of the greenhouse gases yet chlorofluorocarbons (CFCs) that contribute to only 11-17 per cent are said to be much more potent in global-warming than carbondioxide. More than forests, oceans that occupy three-fourths of the globe play such a dynamic role in the total atmospheric carbondioxide flux that oceans and the green phytoplankton need to be protected on priority. Estimates of carbondioxide trapped in polar ice and in pollen grains (palynology) preserved in ice and in sediments since the Quarternary Era show that it is only since the comparatively recent industrial age that the carbondioxide accumulation in our environment has started escalating beyond proportions. A mere doubling of the ambient carbondioxide levels will enhance the severity of the tropical cyclones and storm surges (tidal waves) in the Bay of Bengal by nearly 60 per cent. The Indian space technology and the spacebased observations through satellites and remote-sensing, have such a versatility and capability that we could measure with great precision not only the changing patterns of our natural resources but also the climatological changes including the impending natural calamities.

Natural chemical cycles of the basic elements of our biosphere like carbon, nitrogen, hydrogen, oxygen, sulphur and phosphorus need to be so monitored sustainably as to promote life-supporting systems and their biodiversity. Nearly 25 per cent of the existing biota on earth may vanish by about 2015 AD and hence preservation and promotion of biodiversity should be the key scientific mission on earth.

Tree-farming would be so vital in future. The Second Green Revolution is likely to come through trees and shrubs rather than through agricultural crops.

Climatic changes will have such a profound impact, on all living system, their productivity and trade, that there is a need to regulate all human activities through sustainable life styles. Every economic activity must be accompanied by an ecological obligation.

ENVIRONMENT IMPACT

All our rivers, for example, are polluted through human interference, leading to an evergrowth (eutrophication) by water hyacinth. The increase in mining for building stone and also mining for zinc and lead, have greatly enhanced the pollution around them. Incessant dynamite blasting for sandstone is not only a source of highly disturbing noise pollution but it also causes perennial vibrations in all the buildings around, and throws off clouds of dust into the air which settle down on all vegetation in the vicinity. Naturally, instances of sillicosis and tuberculosis are not uncommon among citizens, much more so among workers in such mining operations.

An integrated anti-pollution and waste management plan envisages on one side the setting up of thousands of toilets in different localities for the slum dwellers and on the other, waste processing plants to convert garbage into pellets which can be sold to industry as alternative fuel. The revenue thus generated could help maintain the project. But this has problems of cost of land availability, maintenance of toilets, their hygienic sanitation, water storage and recycling.

Water Shortage : At present, an estimated 63 per cent of city slum dwellers do not have access to running water. About three litres of water are needed for each use of a flush toilet. If thousands of toilets are built with flushing facilities, how much would be the resultant demand for water, even if a flush that uses less water is devised?

As indsutries are largely responsible for environment damage, it should be made compulsory for all manufacturing units to annually assess the pollution caused by them by way of emission and effluent discharge. It should be mandatory for all units to make an environmental audit.

A mix of legislation and regulation, fiscal incentives, voluntary agreements can create environment consciousness which will prevent deterioration of the environment. The emphasis should be on new units adopting clean environment-friendly technologies than on cleaning up technologies.

The medium and large scale industrial units are to be made fully responsible and accountable for the waste they produce, new units are to be encouraged to go in for clean technologies, effluent treatment plants are to be set up for clusters of small industrial units wherever possible, encouragement to be given to recycling of all waste products, excise and customs rebates for pollution control enquipment are to be reviewed. Besides, the cooperation of industry is to be sought for sound environment management.

A long-term policy for pesticides will incude the introduction of bio-pesticides and non-persistent biodegradable pesticides. A new fertilizer policy would also have to be formulated. The policy should also stress the fact that forests and all vegetation play an important role in cleaning up the air and conserving and regenerating the soil.

Economics had largely tended to ignore the environmental costs and if at all they figured in the accounts, they were grossly under-valued. An attempt should be made to change this. In all aspects of economic activity, the impact on environment should not be ignored.

Reducing emission levels of chlorofluorocarbons (CFC) that damage the earth's protective ozone layer is imperative.

Emissions Quota

A global environmental regeneration programme must be linked to the needs of the developing countries. A large number of the poor of the world live in ecologically degraded regions and they could be employed to regenerate the environment through afforestation, soil conservation programmes, water conservation and grasslands development.

More importantly, the emissions quota must be fixed on a per capita basis and the developing countries emitting less than their quota should be allowed to trade off their unused emissions even as the developed countries reduce their emissions in phases.

Agriculture is a contributor to global pollution as well as the sufferer of the consequences.

The pollution is due to emissions of methane, nitrous oxide, ammonia, carbondioxide and other gases.

Methane is produced in the rice fields and in the guts of ruminant livestock, nitrous oxide results from the bacterial breakdown of nitrogen fertilisers, ammonia is emitted from livestocks and from livestock waste while carbondioxide is produced when vegetation is burnt as part of the cycle of shifting cultivation or when land is cleared for agriculture.

The concentration of these gases in the atmosphere is rising significantly due to industrial and agricultural activities. These gases contribute to acidification, reduction in atmospheric ozone and global warming.

It has been estimated, that the contribution of agriculture to global warming through methane emissions is about seven per cent and about one per cent through nitrous oxide. Approximately half of the global warming results from carbondioxide emissions and about 10 to 25 per cent of these derive from activities related to agriculture.

The developing countries would be seriously jeopardising their future development prospects for years to come if they bind themselves to international agreements on issues related to global warming and biodiversity conservation without ensuring major commitments from the industrialised countries on reduction of carbondioxide emissions and fixation of equal per capita permissible emissions on a global basis.

If, for instance, the developing countries were to agree to forest conservation under the proposed biodiversity convention, they would be conserving a major carbon reservoir which the climate convention also seeks to protect. But are the industrialised countries then prepared to reduce their carbondioxide emission?

There must be a link between bio-resources and biotechnology. Some countries would like to see tropical forests declared a global asset and resource since there is an urgent need to preserve them as sinks for carbondioxide emissions. "But how can forests which are the resources of individual countries become global resources while technologies which can help reduce those emissions remain in private hands?

The main responsibility for the state of the environment today lies with the developed countries which must take corrective action to change their wasteful consumption patterns.

The bottom line is really the additional resources needed for action on a global scale. All such funds would have to be mainly found from the developed countries.

Poverty and development cannot be delinked from environment issues. Environmental standard should be harmonised at the global level.

The developed countries are not only stressing the responsibility of developing countries in the resolution of environment problems of global significance but are also forced to shift a proportionate share of the burden for the protection and regeneration of the environment to the developing countries. This has to be done in the face of increasing scientific evidence that rapid industrialisation leading to increase emissions of carbondioxide and other gases are causing the greenhouse effect.

2. Pollution

AIR POLLUTION

Global air pollution from motor vehicle emissions is growing at an alarming rate because of rapid urbanisation and increasing use of motor transportation. Lax environmental protection laws and poor maintenance of vehicles are exacerbating this situation.

Unless effective measures are taken to curb air pollution, we will be exposed to unhealthy and dangerous levels of air contamination.

The projected increase of the world population from 4,800 million in 1985 to over 6,000 million in 2000 will be largely felt in developing countries. This growth will be concentrated in urban centres where motor transportation will also proportionately increase.

In 1985, 8 of the world's 12 urban centres with a population of 10 million or more were located in developing countries. This number is expected to more than double by the year 2000 to 17 out of 23 urban centres with populations of 10 million or more. In addition, a study says 18 metropolitan areas in developing countries will have populations of between five and ten million.

Atmospheric pollutants associated with motor vehicles—carbon dioxide, nitrous oxide, sulphur dioxide, particulate matter and lead—often exceed World Health Organisation guidelines in many large cities.

Reflecting the fact that automobile ownership is less widespread in Third World than in industrialised countries, ownership in the former has been rising much faster than in the latter in percentage terms. Higher levels of pollution are foreseen for the megacities of the Third World because these motor vehicle growth trends of the Eighties reflect what might be expected into the year 2000, and because of expected continued heavy reliance on poorly maintained buses for public transportation. Compounding the problem of mere numbers, many of the vehicles being driven have no emission-control equipment, such as catalytic converters, and many use low-quality fuel which generates larger amounts of lead and sulfur dioxide.

Urban centres in the Third World—unlike those in the industrialised world—have denser concentrations of people who follow lifestyles that expose them to higher levels of pollution. For example, street space is often limited and polluted air gets trapped between buildings.

Although inspection and maintenance programmes to reduce emissions are an integral element of air pollution control programmes in many industrialised countries, the same is not true for developing countries.

In India, the first stages of a vehicle pollution control programme have been completed. Its major elements include carbondioxide emission limits for gasoline-driven vehicles and smoke density limits for diesel vehicles.

International experience in controlling air pollution from motor vehicles can serve as an important guide for developing countries, the study notes. It suggests a four-part strategy that includes :

Use of Energy Efficient and Clean Vehicles

This requires technological innovations such as catalysts and traps to reduce emissions in new vehicles better fuel management systems such as electronic engine controls; and vehicle inspection and maintenance programmes.

Use of Cleaner Fuels

This calls for reduction or substitution of lead in gasoline and reduction of sulphur in diesel fuel as a first step; then, complete elimination of sulphur, aromatics, and heavy oils in diesel, and lead and other octane boosting additives such as benzene, toluene, and xylene in gasoline, at the same time, it recommends increased use of clean substitute fuels with more alcohol content.

Better Traffic Management

This requires traffic engineering and regulation that will improve traffic flow and manage transport demand.

Overall Policy Framework

This would integrate regulatory, pricing and taxation mechanisms. It would also encourage the use of clean vehicles and fuels and restrict or ban the use of polluting vehicles and fuels.

WATER POLLUTION AND CONTROL

Industrial effluents contain chemical compounds and trace elements which are at a rather high temperature, radio active elements from atomic plants; domestic effluents containing bacteriological contaminants; agricultural runoff containing fertilizers; minerals, salts, herbicides and pesticides; these usually lead to the pollution of water either in groundwater or in surface waters.

The common impurities of water can be listed as follows :

1. *Entrained gases*: Carbondioxide, hydrogen sulfide, methane, oxygen, and nitrogenous and organic compounds.
2. *Dissolved minerals*: Calcium magnesium, sodium, iron and manganese and their carbonates, bicarbonates, hydroxides, chlorides, sulfates, fluorides, nitrates and silicates, alkyl sulfonate from detergents; and synthetic organics from insecticides and pesticides.
3. *Suspended and colloidal materials*: Bacterial, algae, fungi, protozoa, silt and colloidal matter causing water to be come acid in or have colour.

Radioactive Materials: Radioactivity imparted by contact with radiation sources, by entrainment of radioactive substances from mining or processing of overs or by wastes from industrial use of radioactivity.

Water and Disease

The most important of the waterborne diseases are those of the intestinal tract; they include typhoid fever, the paratyphoids, dysentery, infectious hepatitis and cholera. A few parasitic worms may also be disseminated by water. All are due to organisms which are found in the intestinal discharge of patients and carriers and which have found their way into water by some means.

Many waterborne typhoid fever epidemics are caused by the entrance of intestinal discharge from persons having typhoid fever or from carriers of the disease into surface sources of supply, shallow wells or springs and wells. This disease is common particularly in urban areas where untreated surface water supplies or individual wells are used without regard to sources of contamination. Sanitation of water supplies can never be neglected without fatal consequences, water borne epidemics of typhoid fever and more commonly, of dysentery and gastroenteritis. The recognition of waterborne cases of infections hepatitis in recent years may imply that there are many recognized cases.

Certain minerals in drinking water, because of their unknown hazardous properties are considered undersirable if present beyond prescribed limits of concentration. Among these are arsenic, barium, cadmium, chromium, lead, selnium and silver. Anionic constitutents in the same category include fluoride, chloride, sulfate, cyanide and nitrate and there are other harmful inorganic chemicals.

Among minerals not naturally occurring in water and likely to be present through articles means, is lead. The solution of lead in a water with such characteristics may cause lead poisoning.

In many areas groundwater contains fluorides in concentrations sufficient to cause dental fluorosis or mottled enamel. The consumption of water having a high nitrate—nitrogen content or milk formulae using such water is reported to be the cause of "blue babies", a condition known as ideopathic methemoglobining and sometimes causing death of infants.

Waste waters are generally classified as industrial waste or municipal waste. Water characteristics of industrial effluents vary greatly from industry to industry and consequently, treatment processes for industrial waste water also vary. The treatment of municipal and industrial waste can be classified as follows:

(i) Primary treatment

(ii) Secondary treatment

(iii) Advanced waste water treatment or tertiary treatment.

Primary treatment is useful in removing grit, oil and grease and suspended particles where secondary treatment is mainly to remove the organic matter present in the effluents. This can be achieved in two ways, one by aerobic treatment and another by anaerobic method. The unit processes which are involved are activated sludge process, trickling filters, aerobic lagoons etc. Likewise anaerobic treatment processes comprise of anaerobic lagoons, anaerobic digester and anaerobic filter etc.

The tertiary or advance treatment is removal of nitrogen and phosphorous which are responsible for eutrophication of lakes.

Eradication of Guineaworm

The guineaworm disease is caused by the presence of a worm

called "Dracunculus medinensis" in the step wells and drinking water tanks. These worms are swallowed by cyclopes living in water. When these infected cyclopes are swallowed along with water, the cyclopes get killed by digestive system and the guineaworm larvae will grow in the human body and emerge like a thread after a year from any part of human body particularly from legs causing severe pain and handicap.

Control of Fluorosis

The objective is to provide drinking water with permissible limits of fluorides (0.5—1.5 mg/1) to all these affected.

Excess fluoride in drinking water may cause dental and skeleton fluorosis. The control measures include water quality assessment of existing sources, selection of alternate fluoride free source, establishment of treatement plants like defluoridation plants, hand pump defluoridation plants, introduction of domestic defluoridation, conducting awareness camps.

Water Purification

At the International Solar Energy conference, scientists have reported the ways of using sunlight to breakdown harmful compounds such as dioxin, trichloroethylene, pesticides etc. Sunlight concentrated 1,000 times the normal intensity can destroy 99.9999 per cent of the toxic compound dioxin. Special reflectors are used to attain such high intensities. The temperature in the test chamber is around 1000^{0}C where the heat destroyed the vapours of the chemical. This method is not only more efficient than conventional incineration but also leaves fewer byproducts.

In another method, parabolic reflectors were made to track the sun and heat water upto 85^{0}C when hydrogen peroxide mixed with water was found to kill the contaminants. This method is suitable for purifying large volumes of contaminated water. Scientists have used this technique to eliminate trichloroethylene (which pollutes groundwater) and salicylic acid.

CONSERVING WASTE WATER

Water has always been elusive largely due to the fast depleting ground sources. Again, excess rainwater goes waste because of inadequate storage facilities.

A design to conserve this valuable groundwater is suggested for recharging waste water into the ground. The process involves a gully with filter and casing pipes arranged to filter the water (not the type with soil and harmful ingredients) and inject it into the ground by natural means. Excess waste water collected in the gully is expelled to sewers by connected sewerlines.

The body of the device is a gully with or without a drain hole, depending how excess waste water is to be discharged to the sewers or overflown. The base of the gully has a hole to receive the filter unit. While the lower part of the filter moves into the hole, the upper can be seated over the base of the gully by a ring fitted on to the filter pipe. The pipe's top is covered with a removable type cap.

The upper and lower parts are again covered with inner and outer casing pipes with perforated holes. The top of the gully with these arrangements is covered with a grating plate and the outer casing pipe is rested on a soak pit.

Easy Placement

The entire arrangement can be placed at a convenient point within any building. The depth needed is 3" × 3" × 5". After installation, the floor finishing can be completed such that the top level of the gully lies in line with the floor level. During installation the gully is connected to the sewerlines.

Waste water from the kitchen and bathroom goes through the gully and is filtered in stages. The larger sediments settle on the grating plates and the finer ones are retained in the gully chamber. Only water with very fine sediments reaches the filter pipes through the hole on the upper part. These are further filtered in the pipe and near pure water reaches the outer casing pipe and then to the ground through the soak pit.

Excess water reaching the gully chamber is disposed to the sewerlines thus checking overflow. Earth being a good filter, the water with bacterialogical impurities is purified and conserved. Dry wells can also start recollecting conserved water. Any further purification against bacteria, if needed, can be done using existing chemical methods during reuse.

The grating plate, inner casing and filterpipes are separate components, which can be easily removed, cleaned and refitted. This removes any sediments in the parts, ensuring effective functioning of the system.

Excess inflow to the gully will not affect the filtration process as this is discharged to the sewer through the drainhole. Clogging of the soak pit will also not occur. Water conservation is a continuous process. Each time the water-table rises to the level of the gully during the rains, the device will not function and there is no clogging. With excess water, sewerlines are also not clogged for need of minimum flow.

In practice, fairly palatable water used in the kitchen and bathrooms is wasted. But with this process, groundwater can be purified in areas where natural water is saline.

The process can be multiplied as a big unit for larger conservation methods. Filtered water instead of being admitted to the ground can be collected in underground sumps for ready reuse with necessary treatment for bacteria.

There is no maintenance expenditure. The average conservation capacity for clay soil will be about 20 per cent and 70 to 80 per cent for sandy soil. Large scale installations of the system will help canalise water to dried wells.

Waste Water Conservation

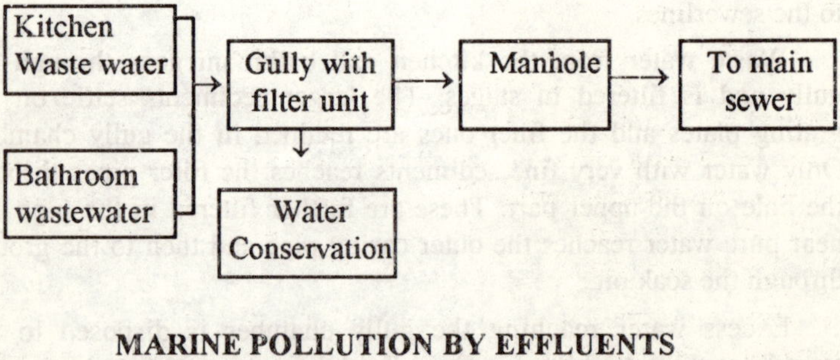

MARINE POLLUTION BY EFFLUENTS

Pollution caused by industrial effluents and other wastes continue to pose a serious danger to the marine life and to the humans consuming some of the species. The waters of important coastal areas of the country, have been affected seriously though in some cases the contents of dangerous pollutants are within the limit or are washed by tidal flushing.

The high oil content accumulates near the harbours due to the discharge of oil and related wastes by the mechanised fishing vessels operating in the vicinity. Concentration of toxic heavy metals occurs in the sea water along the coast and the presence of mercury in the sea and due to the discharge of titanium effluents into the sea. The monitoring studies reveal the effect of discharge of waste water from the soda ash manufacturing industry in coastal waters.

Long term damage: The ammonia entering marine waters where the waste water from a soda ash industry having high pH value and ammonia is released, is oxidised nitrate. Fish mortality, long-term ecological damage and damage of marine life cannot be ruled out in view of the high toxicity of free ammonia. The coastal waters show high levels of pollution during the ebb tide period and this is particularly evident at creeks, where the residence time of waste water is rather high. The pollution of city waters is caused by the huge quantities of domestic and industrial waste waters.

Coast is affected by high contents of heavy metals and other pollutants. Concentration of ammonia due to dumping of municipal and industrial wastes is likely to affect the coastal waters.

Higher levels of DDT reflect the current usage of DDT for vector control in urban localities. The levels of HCH residues show predominant usage of HCH for agricultural purpose. The levels of polychlorinated biphenyls along the coast indicated their usage by the electrical industry.

MARINE POLLUTION MONITORING

The objectives are to : establish a knowledge base in biogeochemical parameters in estuaries, the continental shelf and open seas; operate an appropriately structured information system for ready dissemination of various data to users in government, industry, research and social institutions and to provide advisory and technical services to these agencies for evolving pollution containment parameters; conceptualise and implement R & D programmes that will continually update the knowledge and information bases and develop analytical frameworks for the quantification of transport rates and inputs of various chemical elements to different reservoirs and to the sea towards characterising the eco-system; assessing its digestive capacity and in turn delineating policy options and facilitating decision processes; detect radical changes in the biogeochemical regimes of the oceanic system and to alert government, public and social institutions of their implications; and develop mathematical models both for predicting the dynamics of pollution and ecological regimes along the coasts.

Marine pollution work will help understand the chronic effects of pollution and forecast future risks. Due to the inherent variability of ecological systems at different places, physiological criteria will be taken as an essential component of ecological baseline study.

Irrespective of how sophisticated chemical monitoring techniques become, they will not provide information on the status of the biological components of the ecosystem. Hence a combination of chemical and biological monitoring of the marine environment will be taken into consideration for this pollution monitoring programme.

Pollution potential coastal areas (transects) will be chosen on the basis of location of industries, urban establishments and ecological sensitivity. The work involves collecting seawater samples, sediments and organisms from different transects at different distances and analyse the same, by chemical and biological means.

LETHAL CHEMICALS POLLUTION OF MARINE LIFE

Guarding against the possible effects of poisonous pollutants is warranted.

The deaths of dolphins, seals, basking sharks and other creatures have been blamed on pollution by polychlorinated biphenyl (PCBs), dioxins and trace metals. Concern has also been expressed about the presence of such chemicals in fish.

A comprehensive data base of organic micropollutants and trace metals has to be laid for coastal areas.

Samples are to be collected in selected coastal waters—including the river estuary and the channels—and will then be analysed using the latest techniques, to detect and quantify the many compounds likely to be found in the samples.

Complementary analysis of trace metals in the same samples will also be made. It is possible that mixtures of pollutants are considerably more toxic than the individual contaminants. Scientists are worried about the risks from such lethal "Cocktails", so a key objective will be to use the data in the further development of mathematical models that can predict the dispersion of toxins from the coastline.

COAL POLLUTION

As for coal based power, the quality of coal is a significant environmental issue. The ash content varies from 30 to 50 per cent depending on the source, while the calorific value ranges from 3,000 to 5,500 kilocalories per kg. These vast differences have two undersirable consequences. First, the need to import large quantities of coal and second, the burning of such large quantities results in enormous amounts of flyash (nearly 50,000 tonnes a day) and air emissions that pose a threat to the environment.

According to the Survey of Indian Industry, 1992, coal contributes to over half of the total commercial energy source, dependence may increase with rapid industrial growth, the major consumers being thermal power stations and steel plants.

Global warming, acid precipitation and ozone layer depletion are the three major hazards.

Global climate warming and acid rain are directly relevant to coal of fossil fuel combustion. Increasing emissions of CO_2 and other green-house gases may increase mean global temperature to around 18°C by 2000 (around 15°C in 1990). Nearly two-thirds of the CO_2 released into the atmosphere can be attributed to human activities using fossil fuels, one-thirds is from industrial activity.

The growing acidity in fresh water bodies, forests and soils is due to increasing emissions of SO_2 (by 40 per cent) and NO_x (by 100 per cent) between 1960 and 1990. These pollutants are produced mainly by the combustion of fossil fuels in power plants. They also corrode stone buildings and metal infrastructure.

The demand for more electricity can be met by increasing the levels of power generation from coal-fired thermal stations, bringing with it associated environmental problems.

ENVIRONMENT AND COAL GASIFICATION

Coal can be converted into a gas of medium calorific value of 2500-3000 kcal/nm^3 by oxygen/steam at high pressures. Use of air in place of oxygen gives a low calorific value gas with 1000-1300 kcal/nm^3. The medium calorific value gas can be transported in pipeline and used as fuel while the low calorific value gas can be used locally for heating or power generation. During the gasification process, coal undergoes partial combustion and gasification subject to reaction parameters such as pressure, temperature and velocity of solids in the reactors.

As the temperature in the reactor rises, it undergoes devolatilisation whereby tars, oils, phenols and hydrocarbon gases are released, the char that is left behind reacts with steam and oxygen/air to form several gaeous compounds like carbon monoxide and dioxide, hydrogen and methane, leaving a residue which is mostly mineral matter of ash. Depending upon the maximum temperature attained, the ash is either molten, or agglomerated with partial melting or powdery state. The nature of ash and the gas composition are determined by temperature and pressure of gasification. The granulometry of coal introduced into the reactors determines the design of the reactors.

Environmental Aspects

While selecting an environmentally sound gasification process, several criteria have to be considered depending on the use of the product gases. For integrated gasification combined cycle power generation, these criteria are process efficiency, technological complexities, economic viability and commercialisation and diffusion of the technology.

From an environmental view point, the major areas of concern are :

(i) tars, oils, phenols,

(ii) sulphur,

(iii) NO_x

(iv) particulate matters and

(v) carbon dioxide.

A moving bed gasifier produces tars, oils and phenols which are recovered by quenching the gases with water. The phenols, dissovled in water used for scrubbing the product gases, needs elaborate pollution control methods. The fluidised bed and entrained gasification do not suffer from this deficiency as no tars, oils or phenols remain at higher reaction temperatures.

Sulphur can be removed from the product gas by cooling the gas, adopting the standard processes for gaseous desulphurisation and recovering elemental sulphur. Sulphur can also be removed during gasification stage by injecting a compound such as limestone or dolomite which have a high affinity for sulphur. The sulphur bearing ash is oxidised to convert the calcium sulphide, into calcium sulphate in an external reactor. Otherwise the disposal of calcium sulphide bearing ash can cause environmental pollution by releasing H_2S (hydrogen sulphide). This can be avoided by passing the product gases through a fixed bed of zinc ferrite. The sulphur can be recovered in this case as elemental sulphur.

The formation of NO_x during the combustion of the product gases can be reduced by ensuring sufficient steam injection in the combustor. When the product gases contain large quantity of nitrogen or water as with air/steam gasification, the formulation of NO_x can be reduced below the environmental standards.

One feature of the gasification processes is that most of the ash in the coal ends up as easy to handle bottom slag. The flyash or flyslag is captured by high temperature cyclones and is recirculated for recovering carbon. This ensures that most of the recirculated flyash ends up as bottom slag.

Release of carbon dioxide after combustion of the product gas is inevitable. The integrated gasification combined cycle provides an opportunity to maximise energy generation per tonne of coal consumed. Several steps reduce the release of carbon dioxide per unit of power produced. In the case of high ash coals, benefication to reduce ash level is the first technological option. (It is estimated that for every percentage reduction in ash the useful energy value of the coal increases by 65 Kcal/kg.) Use of coal beneficiated to ash level of 25 per cent would lower carbon dioxide emission by 10 per cent with the current pulverised coal power generation.

In compressed air storage concept, the gas turbine has a higher power output for the same amount of fuel gas. This leads to smaller gasification plant and therefore a lower capital requirement per KW. Compressed air storage is a very effective way to both increase the efficiency and decrease the cost if utilities have adequate caverns for the air storage.

Gas turbines have averaged a gain in firing temperature of 20°C/ year during the past four decades of development. The plant efficiency will continue to increase with the turbine firing temperature increase. There are also developments of more advanced turbine cycles such as HAT (humid air turbine). ISTIG (intercooled, reheat, steam injected gas turbine), and partial oxidation gas turbine.

These advanced cycles will need a higher compression ratio than the industrial turbines. Therefore, current developments of these cycles are based on modifications of the aircraft-derivative gas turbines.

Technological Complexities

Technological complexities of the gasification processes lie in to the pressurised systems in coal feeding, solids movement in the reactor, in bed desulphurisation, ash removal, heat recovery and in gas clean-up. These operations are carried out at high pressures and require proper instrumentation and control systems.

Coal feed to moving bed are less demanding as the only requirement is sized, non-swelling coal. The moisture and ash content have no rigid coal specification in terms of size, and moisture content. It is insensitive to swelling property and ash content of coals. The dry entrained gasification needs very low moisture (less than two per cent) coal while the solid contents in the slurry is a function of coal characteristics. Coal feeding at high pressures weeds lock hopper systems. Pneumatic feeding of dry materials requires good controls which are reduced with screw feeding or top charging.

From technological complexity view point, coal feeding and combustion systems, heat exchange systems, process control, pneumatic coal feeding and fluidisation, ash agglomeration and hot gas clean-up require development of sophisticated and advanced design capabilities.

Technological capability exists in India to design and manufacture systems for coal sizing, moisture control and deashing of coal. Capability also exists for lock hopper systems for coal feeding and ash removal, screw feeding and ash discharge systems.

The control of solids movement in the reactor is fairly complex and knoweldge and experience can only be obtained by actually designing, manufacturing and operating high pressure reactors. Endogenous capability can be developed for in bed desulphurisation using smaller reactors available in India. Product gas clean up and heat recovery systems design and manufacturing capability also exists in India. For hot gas clean-up experience in other countries has shown that this requires high level of expertise in materials technology and it would be prudent to relay on global expertise.

Integrated Partial Coal Gasification : VEW, Germany is experimenting on a combined cycle plant with integrated partial coal gasification in their 10 ton/hour pilot plant. In the partial gasification of coal, over 70 per cent of the coal converted into gas, gasification operating at a working pressure of approximately 20 bas using air as the oxidising agent. The low BTU gas produced is used as fuel for a gas turbine. The residual char from the partial gasification in the gasifier is fed into a modified circulating fluidised bed furnace into which the exhaust gas from the gas turbine is fed to serve as an oxygen carrier for the combustion of the residual char.

Slagging Combustion : The slagging coal combustion is yet another coal burning technology and appears quite promising for Indian coals. In these combustors, coal is burnt in two stage cyclone type combustors to achieve very high temperature and remove most of the ash in the form of liquid slag. The clean gas coming out of the combustor can be used for steam generator, gas turbine or MHD applications.

Magneto Hydro Dynamics (MHD) : The MHD system is based on Faraday's law that if a conductor moves across a magnetic field, electricity is produced in a direction perpendicular to the movement of conductor. Here, the conductor is high temperature combustion plasma at a temperature of around 3000 K. The conduction properties of the plasma are further improved through addition of alkaline salt. High temperature combustion plasma can be achieved either by firing natural gas or coal gas with oxygen as oxidising medium in presence of alkali salts. Coal gas is obtained by gasification processes.

High intensity magnetic fields are used along with high gas velocities in MHD channel to produce D.C. power. The gas temperature is allowed to fall to 2400 K in the MHD channel. Subsequently the steam bottoming plant similar to conventional plants can be employed. By this, fuel to electrical energy conversion efficiency up to 60 per cent can be obtained.

Fuel Cell : Fuel cell is an electro-chemical device, in which two gases, viz., hydrogen and oxygen, are combined to generate electricity and water. While different types of fuel cells are under development worldwide, two are promising at present. These are phosphoric acid fuel cell (PAFC) and molten carbonate fuel cell (MCFC). The hydrogen for fuel cell could be obtained from gasification of coal or reforming of methanol to achieve the maximum energy efficiency, the exhaust from the fuel cell can be used for generating steam which in turn could be used for generating power. This can be very effectively done in case of MCFC as the exhaust temperatures are higher.

THERMAL PLANTS POLLUTION

Coal based thermal stations are located either at the pit head or load centres. In each case there can be serious environmental impact at different stages. Major air pollutants include particulate matter, sulphur and nitrogen oxides, carbon monoxide and hydrocarbons. Often pollutants are diluted in the atmosphere, which is ineffective if the power plants are clustered together.

Suspended particles two to five microns and smaller are harmful. Particulate collectors such as electrostatic precipitators (ESP) or fabric filters can check emission. Nevertheless, a 1000 MW thermal station releases 16 T/D of particulates into the atmosphere.

Acting synergistically with other gases, the suspended particulates aggravate respiratory and cardiovascular diseases, reduce visibility and corrode steel.

Sulphur content of Indian coal is relatively less but with a high ash content calling for more coal to produce equivalent energy. The emissions are, thus, naturally high. SO_2 contribution for a 1000 MW coal fired station is about 1200 T/D using coal with sulphur content of 0.5 per cent. Reduction is attempted through:

— Use of tall stack (upon 275 m) to increase atmospheric dispersion so that ground level concentrations are acceptable.

— Flue gas desulphurisation through regenerative or throwaway wet or dry processes.

— Desulphurisation of coal. The inorganic compound, iron pyrite (FeS_2) is amendable to physical removal by gravity washing. Sulphur in organic form is chemically bound in the coal making its removal more complex through coal gasification, or conversion of coal to a synthetic oil or solid material.

Formation of NO_x, a primary oxide produced in a boiler furnace, is controlled by air-fuel ratio, combustion air temperature, extent of combustion zone cooling and furnace burner configuration. NO_x control from stationary sources is also through flue-gas recirculation. Fluidised bed boilers produce half the NO_x emitted by conventional, boilers firing pulverised coal.

NO_x releases are associated with emphysema, lung diseases, smog, white clothes turning yellow and fading of synthetic fibres.

Other gaseous emissions include carbon monoxide, which is associated with impairment of mental functions, diminished ability and tolerance for exercise. This can be controlled by efficient and complete combustion.

Hydrocarbons cause shedding of tears, coughing, sneezing, headache, bronchitis; acting with NO forms photo-chemical oxidants and smog.

Carbon dioxide and other greenhouse gases are responsible for affecting temperature and rainfall.

Other trace elements like mercury, arsenic and cadmium are present.

Bottom ash and recovered flyash amount to 4184 T/D from a 1000 MW thermal station as schematically shown (see Table, pages 125, 26). Flyash is dumped in low lying areas for reclamation or disposal areas adjacent to the power station. It contains about 60 per cent (S_iO_2) 25.0 per cent (Al_2O_3), 7.5 per cent (Fe_2O_3). The heavy metals present such lead, arsenic, chromium, cadmium, mercury pose environmental problems. Flyash causes water pollution both from indirect and direct flow into the water bodies.

Dry collection of flyash is used to manufacture bricks, cement and aggregates. Guidelines stipulate total use of flyash but, so far, only a small percentage has been put to economic use.

Water Pollution

A lot of water is needed for thermal power station. For a 100 MW power plant, it is 200.000m^3/hr for once through cooling system. Waste water generating streams in a plant include boiler blow down, coal yard runoff, flyash leechets (sailedges), demineralised water treatment and other plant waste.

Studies show that surface and groundwater pollution can occur due to leaching of heavy metals. Trace elements can however, be removed by iron absorption.

Cooling water before release is carried through an open channel keeping the differential temperature below 5^{0}C above the receiving body, to obviate any adverse effect on aquatic life. With diminishing water supply and pollution problems related to thermal stations, there is a need to move towards "Zero Discharge".

Integration of Parameters

For sustainable development it is not only the improvement of skills and methodologies that are required but also the involvement of the public for whose good the development project is conceived. It is necessary to ensure :

(i) Development of an adequate organisational structure to internalise environmental concerns in project planning.

(a) The creation of multi-disciplinary professional groups to integrate the preparation of impact assessment and environmental management plans as integral component of the development project.

(b) Formulation of guidelines for the licensing agencies to assess the capability of prospective lesses to undertake major operations in a systematic and scientific manner.

(c) The creation of an effective monitoring network to ensure that the environmental management plans get implemented.

(ii) Development of skilled and professional managers.

(a) The creation of a cadre of professional environmental managers for the preparation and implementation of environmental management plans.

(b) The development of methodologies specific to the local situation and problems.

(c) The development of norms for the quantification of environmental costs and benefits.

(d) The development of professional groups and centres of excellence dedicated to environmental management problems in the universities and institutions of higher learning.

(iii) Data base and Zoning.

Creation of reliable baseline environmental data to facilitate impact assessment.

Carrying capacity of the following types of zones:

(a) Ecologically sensitive areas.

(b) Areas with a concentration of mining operations.

These concepts have to be implemented since the environmental planning of thermal power stations and environmental management calls for conservation and optimisation of natural resources. (S. Maudgal)

Clean Coal Technologies

Reducing environmental damage is most important. We rely heavily on coal for power generation. This calls for new technologies that can reduce carbon dioxide and other greenhouse gas emissions but will compromise on conversion efficiency.

Problems exist on three fronts—increasing energy demand,

enhancing coal-based energy efficiency and contributing to an environmentally cleaner coal related energy system. The immediate options are to:

(i) Exploit technologies available but not employed due to lack of incentives.

(ii) Upgrade existing technology by suitable adjustments.

(iii) Implement sophisticated clean-coal technologies.

Airborne effluents	In tons
Particulates	3×10^3
Sulphur dioxides	11×10^4
Nitrogen oxides	2.7×10^4
Carbon monoxide	2×10^3
Hydrocarbons	400
Liquid effluents	
Organic material	66.2
Sulphuric acid	82.5
Chloride	26.3
Phosphate	41.7
Boron	331
Suspended solids	497
Solid wastes	
Bottom ash + recovered flyash	3.6×10^5

Annual Effluent Production in Tonnes Without Disulphurisation

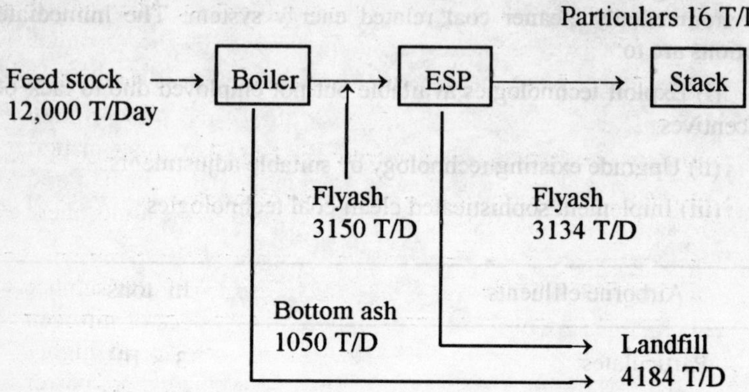

Coal Based Thermal Stations

Stages	Impacts mediated through: Air	Land	Water
1. Coal extraction	Dust, Gases, Occupational health hazards	Soil erosion and Land degradation	Acid mine drainage Siltation
2. Transport & Processing	Dust, Noise, Diesel Exhaust gases	Change in land use	
3. Combustion	Particulates gases (NO_x, SO_x, CO_x)	Combustion products fall out including acid rain Reduction in biomass Productivity	Thermal discharge combustion products fallout including acid rain, Reduced aquatic productivity
4. Ash disposal	Dust	Change in land use	Leechets (especially toxic trace metals and radionuclides)

HAZARDS OF ACID RAIN

The term acid rain refers to rainfall that is significantly more acid than that in remote temperate areas of the world. However, there seems to be a notion that acid rain includes the whole range of processes involved in the emission, transport, deposition and effects of air pollutants, particularly those involving sulphur and nitrogen compounds and ozone. The term acid deposition more accurately describes these.

The constituents of acid deposition can be divided into primary and secondary categories, each consisting of gases and particulate matter. Primary pollutants are produced directly from industrial, vehicular and domestic activities. Secondary pollutants are created in the atmosphere by chemical reactions involving primary pollutants.

Sulphur dioxide (SO_2) is produced during the combustion of sulphur rich fossil fuels and smelting of sulphur rich ores. Oxides of nitrogen are formed mainly by oxidation of atmospheric nitrogen at higher temperature during combustion, particularly in vehicle and power stations.

Emissions of ammonia, (NH_3) are primarily from agricultural sources, though localised emissions around nitrogen fertilizer factories do occur. Hydrogen fluoride is mainly emitted by brickworks, aluminium ore smelting and phosphate fertiliser factories.

The main precursors of acid rain, sulphur dioxide and oxides of nitrogen (NO_x) have both natural and man-made sources. Natural ones include biogenic emissions from terrestrial, tidal and nutrient suboceanic areas and non-biogenic emissions, from natural combustion, geothermal activity, lightening and air-borne soil and water aerosols.

Although natural emissions of SO_2 and NO_x are small compared with man-made emissions, globally they are significant. Natural emissions of sulphur are greatest in the tropics, whereas more than two-thirds of man-made emissions come from Europe and North America. Natural sources of NO account for about a third of global emissions.

Gaseous and particulate pollutants are directly transferred by dry deposition from the atmosphere to vegetation, soils and buildings. Alternatively, they may find their way through rain (so forming acid rain), snow or cloud and fog droplets. They are then transferred on to the surface indirectly under the influence of gravity, as wet deposition, or by impaction as cloud water deposition.

Emissions of primary pollutants are not uniformly distributed across regions. Areas of large concentration occur near the source. The pollutants may be deposited on the ground near the source or chemically transformed in the atmosphere and removed from the area by wind.

Thus the same emission may be a potential threat to crops in industrial areas as gaseous dry deposition, or as wet deposited pollutants in rainfall 100 km downwind of the source.

Air pollutants can affect plants directly through the leaves and stems or indirectly through absorption by the roots from acidified soil. They may act singly or in mixtures and can interact with other plant stresses such as frost, drought, or pathogens.

In temperate areas where average rainfall acidity is about pH 5.0, acidic rainfall (pH 4.0) has occurred throughout Europe and eastern North America over the past two decades. In the tropical rain forests, rainfall acidity may be high owing to the presence of organic acids and in regions with basic soil material exposed to wind erosion, such as Northern China and parts of India pH values regularly exceed six.

In tropical countries, day temperatures and sunlight intensity average high all through the year. As a result production of photochemical oxidants in the atmosphere is very efficient. Seasonal rainfall is also high, leading to wet deposition of sulphur and nitrogen compounds.

In the arid regions, owing to steady winds with little change in direction over prolonged periods, areas downwind of sources may receive high concentrations and deposition of pollutants over long periods.

Gaseous Pollutants	Particulate Pollutants
PRIMARY	
Sulphur dioxide (SO_2) Nitric oxide (NO)] NO_x Nitrogen dioxide (NO_2)] Hydrocarbons Ammonia (NH_3) Carbon dioxide (CO_2)	Fuel ash Metallic particles
SECONDARY	
NO_2 from oxidation of NO Ozone (O_3) and other photochemical oxidants formed in lower atmosphere by action of sunlight in mixtures of NO_x and hydrocarbons.	Sulphuric (H_2SO_4) and nitric (HNO_3) acids formed by oxidation of SO_2 and NO_x. The reaction products (NH_4SO_4, NH_4NO_3, etc.) of sulphuric and nitric acid with other atmospheric constituents, notably NH_3.

FURTHER DANGERS OF ACID RAIN

Acid rain has devastated much of the Northern hemisphere in the last three decades and is now spreading to the industrial regions of the South.

Acid rain spares nothing. What has taken humankind decades to build and nature millenia to evolve is being impoverished and destroyed in a matter of a few years—a mere blink in geological time. (Don Hinrichsen)

Acid rain has been defined as "the fallout of industrial pollutants, sometimes as acidified natural rainfall and sometimes as dry deposition"; or, more specifically, as "rain that is unnaturally acid as a result of pollution of the atmosphere with oxides of sulphur and nitrogen from the burning of coal and oil". It is a versatile destroyer with a planetary implication: it has scourged much of the Northern hemisphere in the last 30 years and is now spreading to the industrialised regions of the South. It lays waste whole forests, damages trees, plants, and crops; acidified freshwaters like lakes, rivers, reservoirs, springs, wells, and subsoil water; induces respiratory illness in humans; poisons wildlife; and inflicts irreversible damage on buildings to deface some of civilisation's most treasured monuments.

What turns ordinary rain, which is very slightly acidic as to be of no great consequence, into corrosive 'acid rain'? The transformation is effected when certain substances released from industrial processes and motor vehicles primarily combine with the water vapour in the atmosphere, water droplets in the clouds and sunlight to produce lethal water-acid mixtures. The release of gases, smoke, ash, sand, and dust into the air has always been there through natural action from wind, volcanoes, forest fires, decaying organic matter and other biological processes, desertification etc. But from the start of the industrial revolution in mid-18th century, escalating human activities leading to increased energy production, new industrial processes for manufactured products and new and accelerated forms of transport, began to influence the ecological balance of the atmosphere by throwing up ever-increasing quantities of constaminants into it. Initially, visible damaging effects such as fog, smog and grime on buildings were tried to be countered by pumping the emissions higher up into the air and dilute them. This, however, simply moved the polluted air to more distant places by wind action; worse, high in the atmosphere the pollutants reacted with water and light and formed a variety of different acids.

These water-acid mixtures fell to earth as rain, snow, hail, dew and fog. New kinds of injury were observed: damaged forests and trees, impaired crops, poisoned aquatic life in freshwaters, and devastated stonework on historic buildings. Collectively, the new plague received the title 'acid rain' which was exactly what it was. It was so recognised by scientists but it was a long time before the governments of the most offending countries would concede that their industries were responsible. And such of course, was the attitude of the industrial bosses. Some still trundle out the same old, tired arguments of the need for economic growth rendering the bad practices imperative.

We can now look at the main pollutants causing the destruction. There are two; sulphur oxides and nitrogen oxides. When coal began to be widely used for power and heat generation and in industry and for heating homes, pollution from its burning became serious. Most coal contains sulphur, and when it is burnt, sulphur oxides are formed. When fuel oil came into use in oil-burning power stations, the same happened. The sulphur oxides reacted with the water vapour in the atmosphere and formed acids including sulphuric acid. Then as consumption of petroleum and gaseous fuels in motor vehicles began to increase massively the last half-century, new pollutants were released in the fumes in their exhausts, and among these were nitrogen oxides. These too reacted with atmospheric water and formed acids including nitric acid. Other emission sources were ships diesel oils (sulphur oxides) and nitrogeneous fertilizers and fired forests and grasslands (nitrogen oxides). Further, other industries released hydrogen chloride which produced yet another acid, hydrogen chloride which produced yet another acid, hydrochloric acid. This is the basic story of acid rain. There are many variations in the chemicals reactions.

Scale of the emissions are staggering. They are greatest, as may be expected, in the industrialised countries with large car-owning populations. Some 150 million tonnes of sulphur oxides and 140 million tonnes of nitrogen oxides are currently estimated to escape into the atmosphere from human action, in both cases approximately equal to the global emissions from all natural sources. (The proportion is nearly nine to one in the industrialised countries.) A single nickel-copper smelting plant in Sudbury, Ontario, Canada, spews out 650,000 tonnes of sulphur oxides each year. Two-thirds of the acid rain problem is attributed to sulphuric acid, nearly a third to nitric acid, and the remainder to hydrochloric acid.

When does rain become acid rain? Acid rain has a pH of less than 5. Recent measurements in central Europe, Scandinavia, the U.S., and Japan have shown pH values of below 5 or even 4 over large areas, while values as low as 2.1 are known in north-eastern U.S. and 2.4 in Scotland.

Some examples of damage to the natural environment and historic buildings will show how long the problem has been allowed to go on unchecked. In Germany, Switzerland and Sweden, half the natural forests are dead or dying-over 60 per cent of all trees in Britain show signs of decay, 80 per cent of all lakes in Norway and a quarter of those in Sweden are acidified and many voided of fish and aquatic life, and 20 per cent of the U.S.'s lakes are sterile.

Damage to the world's most treasured historic and cultural buildings is no less drastic.

Destruction does not have to come in liquid form alone. Dry particles of the pollutants are equally effective. This is often the case in the vicinity of the offending emissions when the sulphur and nitrogen oxides are converted to sulphate or nitrate or to their gas phases and return to earth as particles or gases to make dry depositions. So, whether the pollutants undergo a wet or dry transformation in the atmosphere, when their products return to earth, both are lethal.

The processes by which the various kinds of damage from acid rain occur, the degree of damage, and the influence of other factors in all these, are complex.

Percolation of the acid rain to the water table and subsoil water increases the acidity of groundwater. This leeches (or removes) the nutrients in it which are necessary for plant growth. It also increases the solubility of aluminium, mercury, lead, and other metals which are toxic to growth. The overall result is damage to foliage, stunted root growth, increased susceptibility to disease, and abortive germination. Plants are more sensitive to this kind of attack than animals, but fish and aquatic life suffer greatly from concentration of these metals and are eventually killed off. Fish species like trout and salmon have disappeared from many lakes and rivers in Scandinavia. Animal respiratory systems are not capable of dealing with air contaminated with the sulphur and nitrogen gases, and their further dissolution in the body causes many respiratory diseases.

Nitrogen oxides, besides producing nitric acid, also react with other impurities in the air to make a mixture which has been called "photochemical smog". This penetrates the lungs and interferes with the passage of oxygen into the blood. Sulphuric acid in particular damages limestone or marble (rock mainly composed of calcium carbonate) buildings by reacting with it to form calcium sulphate and carbon dioxide, and the surface peels away or dissolves in the rain. Limestone is excellent for carving and widely used because of this, and has carvings and statuaries destroyed in this way.

There are no easy solutions to acid rain. It is, as we have seen, a result of a mixture of many human activities, and the precise contribution from each is not easily quantified. It is complicated by the fact that the offending community is not necessarily the worst sufferer as the pollutants may be carried hundreds or even thousands of miles and cause blight in distant places. For example, most of the sulphur dioxide emitted by British power stations is exported to Northern Europe. Technology now exists, however, for limiting emissions from power stations, factories, and motor vehicles. But it is expensive and unpopular politically because it adds to energy production costs and car manufacture costs.

Many power stations have turned to low sulphur coals and are gradually introducing emission control devices. The filtering system commonly used is the sulphur scrubber or flue gas desulphurisation vent. This uses calcium hydroxide which is obtained from limestone (unfortunately carbon dioxide, the main greenhouse gas, is liberated in the process).

To protect forests and lakes, emergency measures have to be launched such as spreading lime on them, fertilizing damaged trees with calcium, potassium, magnesium and so on. Motor vehicle emissions are sought to be cleaned by a catalytic converter which is fitted to the vehicle exhaust. This adds to the cost of a new car, but renders harmless almost all of the nitrogen oxides (it however converts the carbon monoxide normally emitted to carbon dioxide, adding to the greenhouse effect). Catalytic converters are now compulsory in the U.S. and Japan.

SOLAR RAYS HAZARDS

Though there is no ozone loss yet over India, areas in south India, nearer the Equator get so much ultraviolet solar rays in summer that they are in a "Virtual ozone hole".

Ozone hole refers to depletion in the ozone layer in the upper atmosphere which blocks harmful ultraviolet (UV) rays from reaching the earth. Man-made pollutants, chiefly chlorofluorocarbons, are held responsible for this loss.

In South India the UV radiation during summer is as high as one expects in Antarctica in September-October when there is noticeable ozone hole over the continent.

"People in South India are already experiencing the UV doses that are causing alarm in Antarctica, but they are accustomed to it."

Indian ozone levels vary between 240 and 350 Dobson units compared to world levels of 110-115 Dobson units recorded during peak ozone hole periods. One Dobson unit equals 0.1 mm of the compressed gas at a pressure of 760 mm mercury and temperature of 0^0 Celsius, referred to as normal pressure temperature.

Components : The UV radiation coming from the sun has three main components—UV-C with a wavelength of 100-280 Nanometers (nms). UV-B with 280-315 nms wavelength and UV-A in the range of 315-400 nms.

All of the UV-C is filtered by the ozone layer and other constituents of the atmosphere. The ozone layer also considerably absorbs UV-B coming to the earth.

UV-B known as sunburn radiation, affects crop growth, causes skin cancers and alters weather patterns, and appears to be most dangerous in the 250-350 nms range.

During summer the local noon value of UV rays at a wavelength of 290 nms at Kodaikanal is comparable to that in Antarctica during low ozone times.

But the overall UV-B radiation (the most harmful) received at the Indian stations is higher than that in Antarctica.

3. Ozone Depletion Hazards

There are holes formed/found in the ozone layer in the earth's stratosphere. The decay of the ozone layer has been of much concern.

The ozone layer over Northern Europe is disappearing twice as fast as previously assumed: Ozone levels in winter and spring had declined by eight per cent in the last decade alone. Ozone loss is not confined to the Antarctic and the Arctic but extends to the rest of the Northern Hemisphere.

There is a grave environmental crisis. Unlike other environmental problems which are basically earth-bound, this is something happening a long way up in the sky.

The make up of the earth's atmosphere is its life-giving gaseous envelope. This envelope which we call 'air' is a mechanical, not chemical mixture of many individual gases (nitrogen and oxygen making up 99 per cent by volume of dry air, with very small or minute quantities of other gases), each with its own physical properties: it also contains water vapour and suspended particles of dust, sea salt, organic matter and smoke. The gas mixture has not always been as it is today, but is the result of a slow evolutionary process from the earth's beginnings.

The vast bulk of it is close to the earth's surface; the gases gradually merge into the emptiness of space becoming virtually non-existent above 100 k.m. The atmosphere is generally divided into four horizontal layers mainly on the basis of temperature: the troposphere (the first 12 km, in which weather phenomena occur), the stratosphere (12-50 km, where 95 per cent of the world's atmospheric ozone is concentrated), the mesosphere (50-80 km), and the thermosphere (above 80 km, the atmospheric density after 100 km becoming extremely low). The boundaries between the layers are not strict, and are known respectively as the tropopause, stratospause, and mesopause.

OZONE AND HOW IT HAS FORMED?

Ozone is a gas, a form of oxygen. In chemical terms, it is an allotrope of oxygen (that is the element exists in more than one form in the physical, in this case gaseous state). Oxygen has two allotropes; the familiar and stable, oxygen (with two atoms of oxygen and represented by O_2), and the unstable ozone (with three atoms of oxygen, O_3). Ozone occurs at ground level (one part in 100 million) where it is man-made (for example, from vehicle exhausts) and is toxic to animals and plants, and it occurs naturally in the upper atmosphere. We are here concerned with the latter. Oxygen molecules in these layers, 80-100 km are split by shortwave ultraviolet radiation (UVR) from the sun into single oxygen atoms, O. These separated atoms collide with other oxygen molecules (O_2) at 50 km and below where the atmosphere is dense enough to make it possible, to produce O_3 (O plus O_2). Below 15 km, the UVR has already been absorbed, and therefore there is little formation of ozone.

Ozone layer. The ozone-rich zone between 10 km and 50 km, in the stratosphere, forms the ozone layer: the peak concentration is around 20-25 km. Even here, the dilution is only 10 parts per million, and it is but a sparse mist. Nevertheless, this layer of the gas is of vital import to life on earth, by providing an obstacle to the burning UVR and stopping an excessive amount of it from reaching the earth's surface. If ozone did not filter a great deal of the UVR, and the UVR reached the earth, life as we know it would be impossible.

UVR. Briefly, it is a part of the electromagnetic radiation emitted by the sun which penetrates the earth's atmosphere. Different parts have different penetrative characteristics and among the materials UVR penetrates are the cells of living organisms, but only weakly so that its energy is absorbed and does not emerge out again. This absorption is beneficial to animals and plants in small doses: it activates vitamin D in human skin, for example. But an overdose can make a break in DNA (deoxyribo nucleic acid), genetic material of most organisms and change cell behaviour, destroying them or making them malignant. Among disorders associated with this and other adverse effects are sunburn, skin cancer, cataract, eye complaints, and immune system suppression. UVR can also slow down plant photosynthesis (process in which energy from sunlight is used to build up complex substances from carbon dioxide and water), which in turn stunts plant growth, impairs seed quality, and increases susceptibility to disease and pests. UVR also destroys plankton in the seas, which could be disastrous to ocean ecosystems.

Although a form of oxygen, ozone has very different properties which are critical to its role of screening out UVR, it's constitution from three oxygen atoms makes it unstable, and thus receptive to breakdown by the energy supplied by the UVR: In breaking down, it absorbs the UVR, acts as a barrier to its further progress earthward, and prevents it from reaching the earth. Thus, when the hours of sunlight are long, the ozone layer is most protective by receiving plenty of energy and absorbing the UVR.

Ozone layer absorbs the energy of the UVR and stops its penetration to the earth's surface.

The ozone layer as a whole is decaying everywhere.

This destructive process has been wholly man-inspired. The major cause is the group of chemicals known as chlorofluorocarbons, or CFCs (reference was made to these in the global warming context), trade name 'Freons'. Their molecules are compounded from chlorine, fluorine and carbon atoms (with or without the addition of hydrogen). They are chemically stable, and released into the air from use as propellants in aerosol cans or from pumping into plastics to make foam, are non-toxic and harmless to life. But when the molecules drift slowly upward, taking years to reach the stratosphere, their effect is quite different. Another, but lesser, cause of ozone depletion is the group of chemicals known as 'halons' (similar to the CFCs, most containing bromine which is even worse than the chlorine in the CFCs at destroying ozone), used in fire extinguishers: the halons do not breakdown till they reach the ozone layer.

The ozone destruction by CFCs occurs thus :

When a CFC molecule is exposed to strong UVR in the stratosphere, it is broken apart. This releases a chlorine atom, which attacks an ozone molecule, pulls away one of its three oxygen atoms, and forms a chlorine monoxide molecule: The ozone molecule is thus destroyed. The destructive process continues further. The oxygen atom from the newly formed chlorine monoxide molecule is pulled away by a free oxygen atom, freeing the chlorine atom to restart the cycle, thus, more and more ozone molecules are decomposed, and the ozone layer becomes depleted.

CFCs are used worldwide as propellants in aerosol sprays, and in refrigerants, coolants, cleaning agents, insulants, and plastic foam. Over a million tons are produced each year, and with this wide range of use, their eventual elimination which is what is needed, is difficult.

Estimates of Ozone depletion. CFCs are not the only agents releasing chlorine in the stratosphere: there are others such as methyl chloride from burning vegetation and chloromethane from rotting fungi. It was also observed that the 'hole', which was first detected at the end of the Antarctic winter filled again soon after: only to reappear in subsequent years, with a depletion rate as high as 90 per cent. CFCs were the main culprits providing the chlorine to destroy the ozone.

So, with the decay of the ozone layer, more UVR comes through the once-effective protective shield. Already, more people today are dying from UV-induced skin cancer (more than 100,000 a year) than ever before, and more people are suffering from eye disorders. There is correlation, a decrease of one per cent in the ozone triggering off a two per cent increase in skin cancer.

The damage to the ozone layer is potentially dangerous.

(If there were no more CFC emissions from today, the chemical reactions from the CFCs already released and which have yet to reach the stratosphere would continue for another 100 years.) Reduction of the release of certain types of CFCs by 20 per cent by 1994 and a further 30 per cent by the end of the century, and also to freeze out the production of halons ware agreed (CFC-driven sprays were banned unilaterally in the US from 1978). The EEC countries have agreed to go for a total CFC ban by the year 2000.

The demand for refrigerants is likely to be an obstacle. New products to replace the CFCs are being developed and some are already available, and may eventually meet demands if toxicity tests are good.

OZONE POLLUTION DISEASES

Ozone pollution levels can double a person's sensitivity to allergens that cause asthma attacks. Ozone pollution at currently acceptable levels is a threat to human health.

Studies explain the recent rise in hospital admissions and deaths from breathing disorders.

Ozone pollution is formed at ground level when hydrocarbons from motor vehicle exhaust and nitrogen oxides from industrial emissions react chemically in sunlight. The pollutant is distinct from ozone in the upper atmosphere, which is formed naturally and shields the earth from harmful ultraviolet radiation from the sun.

The frequency with which people are affected by asthma has nearly tripled. For children under 15, the rate rose 43 per cent.

Admissions for asthma are more prevalent in the summer months. That is when high ozone levels coincide with the peak season for airborne allergens. This suggests a link between the two.

When patients inhaled air with 0.12 parts per million of ozone, they became twice as sensitive to ragweed or grass as they did when they inhaled clean air.

The mechanism by which ozone aggravates asthma might be due to ozone's inflammatory effects on the airways.

It indicates that ozone pollution enhances the intensity of asthmatic reactions.

The effects were seen at ozone concentrations found in cities during summer months.

Stricter standards sought

A peak ozone level in Toronto was 0.15 parts per million and in comparison, New York City reached a high of 0.18 parts per million.

New York has experienced very high pollen counts because of the wet spring. And ragweed, comes in the air, having been blown into the city from outlying areas.

Ozone affects the airways, but most of them involve people who were exercising and breathing much higher concentrations of the pollutant.

OZONE HOLES (MEASUREMENT)

The life-protecting ozone shield in the high atmosphere has weakened by, up to five per cent under the influence of industrial chemicals. That conclusion, based on direct satellite measurements, was announced that many more are now likely to die of skin cancer because of the higher doses of ultraviolet radiation that the weakened shield will let through.

How much worse will it get?

The most important industrial chemicals that chew up the ozone layer are known as chlorofluorocarbons or CFCs. Used ubiquitously as solvents and to transfer heat in air conditioners and refrigerators, the CFCs take years to waft up from ground level to the height of the ozone layer.

Although many years of CFCs are thus in the pipeline, the end of the steady increase in destructive chemicals that reach the high atmosphere may be in sight. Concentrations in the atmosphere will continue to grow until the end of the century, reaching a peak 12 to 30 per cent above present levels, scientists say, and then start to subside. Once fewer CFCs reach the ozone layer, the shield should start to recover, regaining full strength after a century or so.

It is uncertain how much more damage to the ozone layer will result before the new controls take effect. Earlier attempts to forecast ozone depletion, based on mathematical models, were proved so inaccurate by events.

Despite the uncertainty, or perhaps because of it, health authorities are re-emphasising the need for protection from the sun. "There's no question that the ozone problem is really a problem that's not going to go away." One should take sun exposure seriously, particularly in the middle of the day.

Ultraviolet radiation of a particular frequency, called UV-B, causes skin cancer and eye cataracts, and impairs the human immune system as well. Biologists fear that higher fluxes of UV-B would also at some point damage plants, crops and the phytoplankton in the ocean that are the basis of ocean food chins.

In addition to the Antarctic ozone hole, the ozone layer had been depleted in many other parts of the world. At the mid-latitudes the loss over the last decade was twice as great as had been suspected. The declines, measured in the late fall, winter and early spring, amounted to 4.5 to five per cent over the decade.

The same findings applied to the entire Northern Hemisphere temperate zone, with depletions growing more severe the farther one proceeds toward the North Pole. Even larger depletions were found in the Southern Hemisphere.

The findings were based on measurements made by special instruments aboard satellites, and they were consistent with direct measurements of UV-B increases detected by different instruments at widely scattered ground stations in the Northern Hemisphere.

There is more sunlight in the tropics naturally, UV-B levels remain higher than in the mid-latitudes, even with greater ozone depletion in the mid-latitudes. Likewise, summer time UV-B exposure remains higher than that in the winter, even though ozone is more depleted in the winter.

People and other organisms in the mid-latitudes are experiencing unaccustomed UV-B levels. "Organisms in the mid-latitudes have not adapted themselves."

Molecules of ozone-destroying chemicals, when released, spend most of their time up to 100 years in the case of CFCs — below the stratosphere. When a molecule reaches the stratosphere, it gets zapped by ultraviolet radiation, and chlorine is released, and that atom eats up 100,000 molecules of ozone.

Eventually, it degrades, and that ends the process.

Concentrations of another ozone-destroying chemical, methyl chloride, have remained constant, and it is believed a natural constituent of the atmosphere. Including methyl chloride, atmospheric concentrations of the destructive chemicals have increased to four parts per billion from 0.8 part per billion, or just above natural levels.

An additional rise of 0.5 parts per billion by around the turn of the century would be an increase of 12.5 per cent.

OZONE LAYER WARNING

The hole in the Antarctic ozone layer can repair itself but only if there is a worldwide phase-out of nearly all emissions of chlorofluorocarbons and halon gases.

Atmospheric chlorine concentrations have increased to more than three parts per billion from 0.6 parts per billion a century ago. Recovery of the Antarctic ozone hole would not occur until chlorine levels fall below two parts per billion. "Ideally, we should cease emissions of FCCs and other halocarbons immediately".

Options that would not substantially enhance the risk of greater ozone depletions, include substituting hydrochlorofluorocarbons, which leave the atmosphere within 20 years, for chlorofluorocarbons, which can remain for as long as 400 years. "It is a far lesser evil to substitute HCFs for CFCs in order to get global compliance in eventually eliminating these compounds entirely.

Even if all halocarbons were eliminated by the year 2000, it would still take until 2073 for the chlorine levels in the atmosphere to drop to safe levels, and each year's delay will push the recovery date 3.6 years into the future.

THE OZONE LAYER — A GLOBAL POLICY CHALLENGE

With the onset of 'global environmental issues'', environmental protection has assumed significance with respect to the formulation of policy in many countries. The requirement for interstate cooperation to protect the global commons has increased—be it the atmosphere, the seas, space, or increasingly, even the local environment deemed to be of importance to the international community. It may be useful to look at some of the international dimensions of environmental protection as illustrated in the case of ozone depletion, and the lessons we should learn from these.

The term refers to the depletion of the ozone layer in the upper atmosphere, which shelters the biosphere from the effects of excessive ultraviolet radiation. It is caused mainly by the action of man-made gases, known as chlorofluorocarbons or CFCs, which are released into the atmosphere. Ozone depletion has been described as a global environmental issue because its adverse consequences will affect the entire world (though not in the same degree everywhere) and its resolution requires the cooperative effort of all states. The latter is easier said than done.

The consequences of ozone depletion are not the same everywhere: e.g. where greater incidence of skin cancer is worrisome in some countries (as in the temperate climes of many Western states), it is not so in tropical countries like India which have long coexisted with a relatively thinner cover of ozone. This automatically yields different perceptions of the gravity of the problem. In addition, the resolve to solve the problem may not be the same in all states. In India, saving the ozone layer has to compete with other priorities like the provision of safe drinking water to all citizens, or providing primary health-care to all. Questions of equity also come up.

Developing countries understandably resent prohibitions imposed on the present and future use of CFCs. CFCs, after all, have done much to promote development in the industrialised world and raise standards of living and developing countries would like to use them for the same purpose. The fact that substitutes for CFCs reside in the industrialised world, from whom the developing states will have to buy them, does little to increase the developing states' enthusiasm for saving the ozone layer. These problems reveal the tensions which may arise between states over protection of the environment. Depletion of ozone layer is caused by trichloro fluoro methane, dichloro difluoro tetrafluoro ethane, chloro penta fluoro ethane, bromo chloro difluoro methane, bromo trifluoro methane, dibromo tetra fluoro ethane. Trade of these substances with those who are not parties to the Montreal Protocol (with regard to the restrictions) is to be prohibited.

Certain basic principles are discernible in the policy relating to the ozone issue—(1) the polluter must pay (2) measures to protect the environment must be related to the 'capacity to pay' and (3) the opportunity to alleviate mass poverty through economic development must not be denied to developing countries. The approach to the various issues in the case of ozone depletion, such as transfer of technology, protection of the intellectual property rights, scientific cooperation, sharing of information, additionality, strengthening of international environmental law etc., can be related to the above three principles (Mukund Govind Rajan).

4. Recycling

WASTE RECYCLING TO CHECK POLLUTION

Waste-control and pollution curbing measures provide a cleaner and safer environment.

Low and Non-Waste Technologies (LNWT), including environmentally sound technologies, clean technologies, waste recycling, residue utilisation and recovery, are being introduced.

City refuse contains plant nutrients such as nitrogen, phosphorous and potash. Compost from city refuse can be used with chemical plant nutrients to improve soil structure and water-retaining-capacity.

For solid-waste recycling, mechanical composting plants are installed.

If all organic matter in city refuse is converted into compost the heat energy made available can satisfy 27 per cent of the domestic energy requirements of the population. In addition, this also eliminates pollution by solid waste.

A large proportion of paper is recycled.

Plastics also are recycled in a number of ways. Plastic footwear is collected by hawkers and sold to small-scale units which remake them after cleaning and grinding the plastic waste to fine size. Cheap pipes are also made out of the plastic bags and bits.

A mill discharging about 500,000 litres of spent wash daily can generate about 15,000 cubic metres of biogas daily, providing fuel for boilers for about seven hours. This process also saves bagasse from being burnt in boilers resulting in reduced fly-ash pollution.

Other recycled products include disposed bottles and newspapers, silver from used photofilm and paper, lub oil, and packing material like gunny bags and cardboard boxes.

The treatment and conversion of urban waste into energy reduces pollution in cities and help in the much-needed energy generation.

Power is generated by burning sludge gas from a sewage treatment plant in life-expired aero engines.

One promising technology developed involves the production of cement from rice husk-ash.

GARBAGE RECYCLING

In India, around three lakh tonnes of garbage is generated every day. The country is already spending Rs. 23 crores annually on disposal of this garbage. Most of it is produced by urban houses and industries. The problem is typically "dump and forget".

While waste recycling — principally of paper, plastics and metal items in garbage—came into vogue in developed countries recently, the ragpickers in India have unobtrusively been performing this function all along. But garbage contains a good deal of cellulosic matter in a form which can be burnt to produce energy.

However, recycling plants producing energy from garbage have been around for a long time. 12,000 tonnes of refuse (fuel) every week generate 68 MW of power.

Indian garbage is generally low in recyclable components (which probably would have been collected by ragpickers) but has much higher moisture content and large quantities of inert materials like sand, grit, stones and small metal scraps. As a result, raw Indian garbage has low calorific value and does not burn easily. Unlike garbage in developed countries, raw Indian garbage cannot be directly incinerated to generate power.

To increase the garbage's calorific value before using it as fuel it had to be processed. Since its composition is different, the Indian garbage has to processed differently from those in the West where the intention is mainly to separate recyclables.

A continuous process to remove sand, grit and metal pieces from the raw garbage, dry it, shred it, and finally compact it into fuel pellets is to be adopted.

The technology needed is simple. The garbage is passed at various stages over metal meshes to sieve out soil particles. Shredders chop the garbage clumps into small pieces for easier processing. The garbage is passed through a kiln to dry up the moisture with hot air circulation. (It also serves to kill pathogens in the garbage particles.) Iron can be separated using magnets while non-magnetic particles like glass, ceramics, non-magnetic metal pieces and stones are removed by sucking the garbage vertically upwards in a strong stream of air.

Mixing agricultural wastes like coffee husk and sawdust into the processed garbage further increases its calorific value. The mixture is then made into pellets using a modified cattle-feed pelletiser. This increases the calorific value from 937 per kg for the raw garbage to over 4000 kcalories per kg for the blended fuel pellets (with agricultural wastes mixed in).

The plant set up operates eight hours a day, processes 12 to 15 tonnes of garbage every day to produce four tonnes of fuel pellets a day (500 kgs per hour).

A 60 tonne (per day) plant can process about 200 tonnes of raw garbage every day. It has to be in proximity to a garbage dumping yard.

The country's first large-scale plant produces fuel pellets from municipal garbage. The plant is designed to convert 150 tonnes of garbage (Bombay generates, 4,000 tonnes every day) into 80 tonnes of pellets which can be used by local industries. Nearby slum-dwellers too can use it as a cooking fuel.

The garbage is first spread out and sun-dried in an open yard for a day. This alone removes a 30 to 35 per cent of the moisture in the garbage. (Some items like chappals and tyres are picked out by hand. One hectare is enough to dry 250 tonnes of garbage.) Similar procedures are used to process the garbage. The light combustible part of the garbage is separated from the denser fractions by ballistic separation where a conveyor belt drops garbage from a height while a blower provides a stream of air so that the lighter particles fall clear of the heavier ones. A part of the unprocessed garbage, after a day's drying is burnt to run the kiln in which the garbage is dried. Some critical equipment like grinder and pelletiser are run in two streams.

With proper separation of the non-combustible part of the garbage and adequate drying, it is possible to get pellets with a calorific value of 3500kcal per kg. Operating methods are being evolved to cope with various kinds of garbage.

Such plants, should be designed only after carefully examining the kind of garbage available locally (There could be seasonal differences even in the same place). In Mumbai, for instance, there are certain seasons when large quantities of hay (used as packing material) turn up in the garbage, increasing the calorific value. Moreover, garbage from road sweepings is entirely different in character from vegetable market wastes. In addition, climatic changes should also be considered. For instance, the plant could not be operated during the three monsoon months.

Composition of Urban Refuse (Percentage by Weight)

Constituents	Industrialised (New York, 1962)	Middle Income (Manila, 1970)	India (Calcutta, 1975)
Paper	35	17	3
Glass, ceramics	9	5	8
Metals	13	2	1
Plastics	10	4	1
Leather, rubber	-	2	-
Textiles	4	4	4
Wood, bone, straw	3	6	5
Non-food total	74	40	22
Putrescible	22	43	36
Miscellaneous inerts	4	17	42
Compostable total	26	60	78
Per capita garbage generation (kg/day)	0.7-1.8	0.5-0.9	0.3-0.6
Density (kg/cum)	100-150	200-400	250-500
Moisture	20-40	40-60	40-80

Source : Srinivasa Setty and Nagaraja.

The garbage can be gasified at temperatures of 500 to 1000°C in the absence of oxygen. This process (called pyrolysis) reduces the waste to gases comprising hydrogen, methane, carbon monoxide and carbon dioxide, tarry liquids containing wastes and organic compounds (such as acetic acid and methanol), and carbonaceous char. Some of the gas can be burnt to run the reactor itself. The surplus gas and liquids are available as fuel products. The solid char can be processed to yield various grades of useful carbon such as activated carbon or an inert slag which can be disposed off easily.

Alternatively, cellulose in garbage can be converted into glucose enzymatically or through chemical hydrolysis. The glucose can be fermented to produce ethanol. Other useful chemicals like acetone, organic acids, glycerol can also be extracted through the hydrolysis of garbage.

WASTE CLEANUPS BY BACTERIA

A strain of bacterium usually found in mud might be the key to cleaning up waterways and nuclear waste sites contaminated with uranium.

The rod-shaped bacterium, known as GS-15, usually, derives energy in a biochemical reaction that changes iron from one form to another. Laboratory experiments indicate that the microbe has a similar effect on uranium. The microbe transforms the highly soluble radioactive metal, which is difficult to isolate in waste water, into a solid form that can be easily filtered out.

Scientists have shown that microbes are involved in changing uranium from one form to another, but it was thought to be an indirect process. The new study demonstrates that this bacterium metabolises uranium directly, deriving twice as much energy as it normally would be in the presence of iron.

This is a very significant. It helps explain the presence of uranium ore in sediments, but more importantly it offers possible mechanism for cleaning up environments contaminated with soluble uranium at a very rapid rate.

The bacterium was isolated from muddywater. The microbe could be cultured overnight, and its growth could be stimulated with the addition of certain nutrients.

Waste water from uranium mining and at nuclear energy plants is often contaminated not only with soluble radioactive metals, but also, with toxic solvents. GS-15 might disarm both.

Waste could be pumped through a bio-reactor seeded with the bacteria. In such a vessel, the microbes could decompose the toxic compounds and transform the dissolved uranium into an insoluble metal, which could be removed, leaving purified water to be returned to the environment.

RECYCLING OF OLD CONCRETE FOR NEW ROADS

Resurfacing will now make use of concrete which is 30 years and more. The existing concrete can be recycled and used for the new surface. The asphalt layers on top which get mixed with the concrete when the surface is broken up and removed will not impair the quality of the new surface.

Hitherto it was assumed that old concrete mixed with asphalt could not be reused for road surfacing. Concrete made with fragmented surfacing and with old concrete has shown that even if the old concrete contains 20 per cent asphalt, its quality is enhanced by the mixture.

METAL RECYCLING

Added value is an increasingly important factor for all areas of recycling as a rising proportion of feedstocks consists of lightweight, often contaminated materials.

With this in view, processors of nonferrous metals—especially aluminium—have been showing great interest in converting reclaimed metal into specification ingots for direct sale to consumers. But rising costs and tougher environmental controls are putting pressure on a popular area of diversification.

An extremely high recovery rate is achieved by melting the preheated scrap under an controlled flame molten aluminium separating for contaminants and running to the back of the dry hearth chamber, which incorporates a holding bath. Recovery rates are further increased by melting a certain amount of scrap in the liquid bath.

Patented sealing of the furnace doors prevented emission of fumes and heat, and a unique afterburner/recuperator system provides heat recovery. Flue gas temperature is reduced from 1000°C to 400°C. Tests have shown that gases exhausted from the plant are often cleaner than those being drawn into it.

The total system offers a combination of new and existing technology that enables the processing of secondary metal at 35 per cent of what has been regarded as the standard industrial cost. Previously uneconomic material now becomes economic.

Air filtering system is geared to the needs of operators of sloping hearth and reverberatory furnaces of up to five tonnes capacity.

Fundamental Crushing

A car body could be drawn through the crushing stages without recourse to winches or independent handling equipment. In a fully automatic sequence, a hydraulically-driven claw device pulls the shell inwards retracts as the compression is made and in a raking motion, pulls the car forward for each compression stroke and finally ejects it. A standard saloon car body can be compressed within four minutes.

Pyrolysis Unit

A pyrolysis unit converts old tyres into fuels and other reclaimed products.

Disposal of discarded tyres is a serious problem in all Western countries, and there is a limited demand for the rubber that can be reclaimed from them. Their use as a fuel it attractive, but environmental considerations present some costly control problems.

Pyrolysis—essentially chemical decomposition by heat—offers the advantage of processing in a totally closed environment and generates useful fuels in the form of oil and gas.

The unit consists of a refractory lined kiln, a condenser, and a gas scrubber that uses circulating water as its coolant. It is designed to process about 1000 tyres a year and, with throughput increased by batching, it can be suitable for large-scale industrial use. The oil and gas generated could be used as fuel in combined heat and power systems.

Can Sorting System

One of the by-products of increased public interest in environmental and resource conservation issues is the steadily-rising recovery of post-consumer cans whether tinplate or aluminium.

An automatic baling and sorting system comprises two stand alone pieces of equipment. It provides the sorting and conveying system, using magnetism to segregate the tinplate from aluminium cans at a rate of up to one tonne/hour.

Designed to meet the needs of can processing on a larger scale the unit sorts and flattens cans prior to baling or bulk transport to a melting plant. A hopperfed feed conveyor provides a facility for hand-picking to remove foreign matter before the cans pass between counter-rotating rubber surfaced rolls to flatten them. An overband electro-magneto then pulls off tinplate cans, which are deposited via a conveyor to a waiting container.

Aluminium cans are discharged from the machine into a separate container. Output is of the order of 1.5 tonnes/hour and a useful resource is saved from its previous fate of landfilling.

BIOGAS PLANT EFFLUENT USES

Anaerobic fermentation technology of biologically degradable organic wastes not only produces methane-rich biogas, but also nutrient-rich manure.

There are now more than seven lakh family size biogas plants and 300 community biogas plants working satisfactorily. Slurry nutrients can be harvested for production of feed and food and replace conventional inorganic fertilizers partially or wholly.

An analysis of biogas plant feed material (cattle dung) and digested manure (biogas plant effluent) shows that major nutrients such as nitrogen, potassium and phosphorus are recovered almost completely after its production.

Nutrient status of cattle dung

	Cattle dung (brefore biogas production) (in per cent)	Digested manure (after production)
Total Nitrogen	1.20-1.70	1.22-1.72
Ammonia Nitrogen	0.18-0.48	0.26-0.90
Total Phosphorus	0.31-0.57	0.15-0.45
Total Potassium	0.86-1.58	1.14-1.46

The ammonia form of nitrogen increases to 70 per cent during biogas production. As plants and algae prefer this form, fresh slurry can be applied to agriculture fields or fish ponds.

The aquatic plants, *Lemna* (common duck weed) and *Wolffia* (commonly called water meal) use the available form of nutrients from the diluted biogas plant effluent (one per cent) with the yield of 22-39 g of fresh biomass/m^2 per day. These plants contain crude protein upto 39 per cent as against only 10-20 per cent cultivated by conventional methods.They can be used as fish, poultry or cattle feed.

Using biogas plant effluent in aquatic plant production and fish culture has promising results regarding application as an organic fertilizer in aquaculture. Fish culture with organic wastes in an old practice sewage is directly mixed in a fish pond: human waste, animal and agricultural wastes have been used directly or after compost in ponds.

Using biogas plant effluent is the source of fertilizer for fish culture to boost yield. The Tilapia fish (*Oreochromis mossambicus*) is grown.

Tilapia growth in biogas plant effluent

No. of fish per hectare	Initial size (g)	Maximum weight obtained in 125 days (g)	Total production per hectare per 125 days (kg)
30,000	0.5	80.100	2,400
60,000	0.5	73-93	4,400

The only problem at higher density is maintenance of water quality. The *Tilapia* fish reproduce throughout the year.

Polyculture, growing different species of fish in a single pond for better use of feed and fertilizer, is popular among fish farmers. The Indian major crops like Rogu, (*Labeo rohita*), Mrigal (*Cirrhina mrigala*) and the common crop (*Cyprinus carpio*) are grown in 4 : 3 : 3 at a density of 5,000 per hectare. The effluent is mixed at 0.15 per cent concentration every three days. The mixing may be reduced or stopped after three months by checking water quality.

Fish types grown in effluent

Fish	Initial size (g)	Maximum weight attained in one year (g)	Total production/ hectare per year (kg)
Rogu	1.3	1670-1950	
Mrigal	1-2	1210-1410	5,500
Common carp	0.5-2	660-700	

These fish types are grown in biogas plant effluent without any supplementary feed, thereby reducing the cost of additional feed and fertilizer. This way, there is little chance of microbial attack, since, the effluent is free of pathogens. Fish weight increases five to seven times in this effluent than conventional production with recommended doses of chemical fertilizer.

RECYCLING PLASTIC BOTTLES AND RUBBER TYRES

Discarded plastic bottles and rubber tyres are connected into new products such as sound absorbing components for cars.

Current methods of disposing low-grade plastic and rubber waste are based either on incineration, which makes no use at all of the intrinsic properties of the materials, or conversion into low-grade products. By contrast, it is planned for a new process to retain and exploit the plastic and elastic properties of the waste material rather than using it simply as an inert powder.

SEWAGE SLUDGE UTILISATION

Sewage sludge, one of the most intractable environmental problems could be cheaply and easily converted into dried granules for fertilizer or fuel. At present millions of tons of sewage sludge are dumped in the sea.

The sludge is what is left when water authorities remove municipal waste water from the sewage system.

There are other alternatives to sea dumping but they often pose environmental and practical problems. Spreading the liquid sludge on farm-land requires transporting the foul-smelling material in fleets of lorries. Also, sludge from industrialised areas contains heavy metals, which makes it unsuitable for use as fertilizer.

Raising a stink

Sludge can be coagulated and compressed to reduce its volume. The sludge cake can then be spread on farmland. But this can be a labour intensive and smelly operation. Incinerating sludge also causes problems; it generates carbon dioxide which contributes to global warming.

Raw sludge entering the system is mixed with large granules from the end of the system that were too bulky to be used as fertilizer. This thickens the sludge so that it can be heated up more evenly.

The thickened mixture enters a rotating drum along with hot air from a furnace and is dried at temperatures exceeding 450°C, to produce a granular mixture. In the next chamber, hot gases and vapour are drawn off and fed back to the furnace. The granules pass on to a grader which separates out the large granules for recycling.

Because the whole process is a closed circuit, no dust or odours are generated. Also, up to 70 per cent of the energy generated by the furnace is reused and the thermal efficiency is 80 per cent. The granules, which can be stored indefinitely, serve as fertilizers rich in nitrogen and phosphate. Alternatively, water Agencies can sell them as fuel because they burn as well as low-grade brown coals but do far less damage to the environment because they contain less sulphur. If the granules are too rich in heavy metals, they can be buried in landfill sites or incinerated.

The process will work out considerably cheaper than conventional methods. 500 tanker loads of liquid sewage sludge each week could be reduced to just 20 lorry loads.

FLYASH UTILISATION AS BUILDING MATERIAL — A POLLUTION CONTROL MEASURE

Flyash use is a global problem reaching complex dimensions in countries generating power from coal based plants. India too depends largely on coal and lignite fired plants. Both leave 30 to 40 per cent of their weight as waste material in the form of ash which comprises pulverised fuel ash (PFA) called flyash, furnace bottom ash (FBA) and cenospheres.

Being finer fraction, flyash is an important by-product which constitutes 80 to 85 per cent of the total ash produced. FBA clinkers form larger particles, which range from fine sand to 75 mm lumps and form 20 per cent of the total ash generated. A small portion of PFA in the form of hollow or solid microspheres (cenospheres or floaters), five to ten per cent of the total ash by volume, floats to the top in the ash pond.

At present, coal ash is disposed by dumping dry in open areas, mixing with water and discharging the slurry into ash ponds and discharging into nearby ravines, if any. In the absence of immediate users, collection of dry flyash is very expensive and polluting. Thus it is collected in wet form or slurry. Every tonne of flyash needs eight to 10 tonnes of water to convert it into slurry, while every MW installed capacity needs around 0.4 hectare of disposal area.

The 66 big thermal power plants in India are the source of flyash.

About 110 million tonnes of coal used annually here, produce about 30 to 40 million tonnes of ash or one lakh tonnes a day. With 15 per cent annual rise and thermal capacity boost slated for the decade, the production of coal ash is likely to cross 130 million tonnes a year, about 3.5 lakhs a day, by turn of the century. If unused, the disposal will need 28,300 hectares of storage land, a serious ecological problem for land, water and air.

The Indian Flyash

The main mineral species identified in flyash are mullite, tridymite/quartz, magnetite and amorphous glassy phase along with unburnt carbon. As, glassy phase is between 30 to 60 per cent, flyash has a unique pozzolanic property. While chemical and physical properties mainly depend on coal quality, grinding and combustion techniques and boiler operation, flyash contains carbonaceous matter, varying from two to 15 per cent, usually less than ten per cent. The carbon is coke type and coarse.

Use in building materials

Large amounts of flyash can be used in building roads and embankments, reclamation of low lying lands and refuse dumps, land-filling, treatment of polluted waters and unsuitable soils for agriculture, and extraction of some rare metals. While such uses hold great potential of bulk disposal, research and development during the past three decades has realised its use in building materials. Some important ones being—

(i) Portland-Pozzolana cement.

(ii) ready-mixed and precast flyash concrete blocks;

(iii) sintered flyash lightweight aggregate;

(iv) lime flyash cellular concrete;

(v) flyash bricks;

(vi) portland cement clinker;

(vii) oil well cement and

(viii) masonry cement.

Disposal as raw material

Applications in construction may be classified upon its characteristics and availability in dry or wet form. To manufacture cellular concrete, clay-bonded flyash and lime bricks or blocks, dry or wet flyash is used as raw material with lime, clay and gypsum.

Recent investigations on Indian flyash have shown their suitability for use as admixtures for concrete in dry form. It has been found the strength developed in flyash in 28 days normally equals that of corresponding portland cement concrete. Admixture can be prepared on site by mixing portland cement and flyash in a concrete.

In the process, the clay can be replaced to the extent of ten to 30 per cent of flyash. The rest of the process is the same as in manufacture of hand-moulded burnt clay bricks.

Calcium silicate bricks: A variation of sand-lime bricks, it uses flyash instead of quartz sand. The process involves low or high pressure compaction followed by either low pressure steam curing or autoclaving under elevated hydrothermal conditions. The bricks produced under high pressure are superior but need fairly large capital investment on autoclaves.

Portland pozzolana cement: Ten to 25 per cent dry flyash is mixed with clinker during manufacture of cement or blended with finished ordinary portland cement (OPC) to produce portland pozzolana cement (PPC). PPC requires more setting time than OPC but is as strong and can be used in place of OPC under normal conditions.

Being a source of SiO_2, Al_2O_3, flyash is suitable for use instead of clay in the raw-mix for producing portland cement clinker by semi-dry process. As per present estimate around three million tonnes of dry flyash per year can be used for the production of PPC.

Cellular concrete: Light weight aerated or cellular, concrete is manufactured by mixing flyash, quick lime or cement and gypsum in a high speed mixer to form a thin slurry. A small amount of foaming agent such as aluminium powder is mixed into the slurry releasing hydrogen. The aerated slurry is poured into steel moulds and allowed to set. The blocks are then removed after the initial set and autoclaved at steam pressure. They are then allowed to cool and stacked for use.

Flyash lime-gypsum blend (Fal-G) : Phosphogypsum, a waste product from fertilizer plants is washed and refined, followed by calcination. Flyash and lime mix in different proportions and properties with calcined gypsum, to produce Fal-G with strong binding properties and which can be used as cement Fal-G can be mixed with sand or aggregate to produce building blocks of any desired strength.

Utilisation Prospects of Flyash by 2000 AD

Medium Value	Low Value
1. Light weight aggregate	1. Used mine fills
2. Flyash concrete	2. Embankments
3. Pozzolana cement	3. Backfills
4. Cellular concrete	4. Highway base
5. Bricks/blocks	5. Soil stabilisation etc.
6. Grouting	6. Structural fills
7. Slabs and wall panels	7. Water dam concrete
8. Prefabricated building blocks	

PLASTICS RECYCLING

More enlightened retailers are becoming aware of the growing consumer resistance to overpacking and are cutting down on unnecessary plastics. Packagers are under pressure on two fronts. First from consumers who are beginning to resist goods they view as over-packaged. Second, a change in the way many products are packed. Litter problems, diminishing landfill sites and the high cost and possible dangers of domestic waste incineration are becoming acute. Packaging is replaced by returnable containers.

40 to 70 per cent of the containers are cycled — either, burnt as energy or made similar plastic products.

The solution to the problem of domestic plastic waste is nuclear. Environmentalists first argued for the use of degradable plastics that would rot in landfills. But they soon discovered technical problems with this solution and now champion recycling, although experts warn that this too could be dangerous.

Industry, especially the food sector, is confused because decisions on packaging are fundamental to the presentation, transport and shelflife of its products. There is a certain ambivalence among consumers. They want convenience foods but they also want the product to be environment friendly.

Part of the reason for this ambivalence is confusion about plastics and the multitude of materials lumped under the tag. Thermoplastics, which include a number of packaging plastics have become part of our way of life because of their light and long-lasting properties.

Although thermoplastics are recyclable, packaging is often made from a mixture of several types. This makes recycling difficult and often impractical. A bottle made from polyethylene terephthalate (PET), for example, is inexpensive to make and so light in weight that it is cheaper to transport than, say an equivalent glass bottle. Like glass, PET can be recycled as long as a collection system is set up.

A logical, but as yet unrealistic, alternative to the everlasting properties of plastic waste is to use an equivalent that is designed to rot. Most plastic is made mainly from oil. Hydrocarbon molecules, called polymers, are repeated in chains that are too long for microbes to break down. These chains give plastic its unique qualities and make it ideal for packing sensitive products, such as pharmaceuticals and foods.

In the 1970s, researchers produced semidegradable plastics by introducing starch at regular intervals in the polymer chain. When buried, microorganisms feed on the starch and the plastic disintegrates into smaller pieces. This sort of plastic, often used as shopping, post and rubbish bags, is more correctly termed bio-destructible. It is fully biodegradable because it fails to break down entirely into carbon and water.

Bio-destructible plastics take much longer to break down if left as litter. One of the solutions to this problem is to make the polymer chains sensitive to ultra-violet light. Photosensitive plastic breaks down into smaller pieces if left in sunlight but the process will not work effectively if the product is buried. These plastics are up to 15 per cent more expensive than the conventional equivalents but are favoured by some companies to trumpet their environment sensitivities.

Chemical companies are working hard to produce fully biodegradable packaging that retains the characteristics of plastic.

A biodegradable plastic from starch has many of the characteristics of petroleum-based plastics and could be used for disposable cups and plates now made from polystyrene.

A starch-based packing material suitable for salvers, trays, cups, dishes, boxes and egg-boxes is biodegradable.

A starch-based biodegradable material suitable for transparent ''blister-packs'' used to package a wide range of consumer goods including batteries and toys will degrade into carbon dioxide and water when it comes into prolonged contact with water or moist soil.

Soft Plastics

Plastics are being made from vegetable oils. The oil's fatty acids are linked to produce plastic-type polymers. This plastic should, pose no threat to the environment and by-products could be used to reduce fertilizer consumption on farms.

BIODEGRADABLE PLASTICS — I

Plastics pose a serious waste problem in industrialised countries. They can be burnt, but this releases toxic gases into the atmosphere and damages incinerators. Or they can be buried. But, large volumes of plastic wastes take up scarce landfill space and cause problems in land reclamation by making the ground loose and unstable. When dumped at sea, plastic wastes often end up strangling birds and killing marine life. Worst of all, since they are nearly indestructible, plastics never simply disappear. But, what if plastics could be made durable enough for their product applications and degradable over the long-term?

This is the starting point for researchers in biodegradable products such as high-starch filled polymers, biologically derived materials and composting-based materials. Most degradable plastics technologies add an agent to the plastic which causes its disintegration through biological processes (biodegradable) or light (photodegradable).

Biodegradable plastics show great promise, because micro-organic enzymes dissolve them into the soil. Most contain starch as an additive. Micro-organisms in the soil or water consume the starch particles in starch-based biodegradable plastics. This mechanically weakens the plastic and causes it to fragment. An internal catalyst reacts with the polymer causing further break down into sizes that microorganisms can consume.

Most first-generation biodegradable products contain only seven per cent starch. This is the maximum loading that materials such as polyethylene, polyolefin, and polystyrene can have without processing or equipment changes. The catalyst levels also are limited by process and product performance considerations.

A process has been developed that allows up to four times, more catalysts to be added to a standard starch-based biodegradable plastic, thereby speeding up oxidation. This material can be used to make disposable bags for composting applications, where the life of a plastic film needs to be 18 months or less.

The catalyst is essentially a transition metal salt that, when exposed to any kind of temperature, produces free radicals which attack the polymer. Shelf life of these products in much shorter than that of conventional ones and date coding might be required to prevent premature degradation.

Scientists are developing 50 per cent starch filled plastics. They use starch- and cellulose-based synthetic polymers as compatibilisers to blend starch acetate and cellulose acetate with corresponding polymer blends. The resultant material has properties that match those of polyethylene and is only slightly costlier.

Simulations of high-starch loadings showed that high-starch loadings accelerated overall degradation rates. These simulations revealed that a threshold value existed at a 59 per cent starch loading. Below this only 16 per cent of the starch particles were accessible to each other. Above 59 per cent, fully 77 per cent of the starch particles were accessible to each other creating a percolation effect for rapid material degradation.

A commercially available wholly biodegradable plastic was made from renewable resources by the fermentation of sugars.

The new 'Biopol' material can be completely broken down on disposal by the action of naturally-occurring fungi and bacteria into carbondioxide and water in the same way as any other organic matter. In normal use, however, it is comparable in durability, stability and water resistance to ordinary thermoplastics. It can be processed by conventional manufacturing methods to make a wide range of articles such as bottles, films and fibres.

Biopol was chosen to be moulded into bottles for shampoos. The material has a wide variety of other potential applications, including medical uses which can exploit its benign, versatile properties and slow degradation in the body, an agricultural use for the slow release application of pesticides over prolonged periods.

The rate at which the plastic degrades depends on a complex series of factors which influence microbial numbers and growth. Under "favourable" conditions such as those found in anaerobic sewage, it can disappear completely in a few weeks. The material can also be recycled and is particularly suitable for domestic items that find their way into compost systems, soil, water courses, sewers, and landfill waste disposal sites, or for materials intended for agricultural use.

Biopol is the first "natural thermoplastic" material to be made by mocro-organisms in a fermentation process to research a commercial application. It is produced from carbohydrate/organic acid feedstock by the naturally-occurring bacterium *Alcaligenes eutrophus* in the form of long chain molecules of carbon, hydrogen and oxygen as an energy reserve, much as humans store fat. A series of Biopol polymers with different properties of strength, flexibility and toughness can be created by changing the fermentation conditions under which the bacteria are grown.

Degradable polymers often take more than two years to degrade and, in addition to CO_2 and H_2O, produce a humic material consisting of complex organic compounds.

Recycling currently assumes a larger role than degradable materials in solid waste management, but the waste problem is too complex to solve with any one process.

Because different plastics are processed differently, melting down a mixture of milk bottles, diapers, and garbage bags will not result in a useful material. Even the landscaping timbers and outdoor furniture now being made of commingled plastic scrap must contain a substantial amount of virgin material to make an acceptable product.

A new totally biodegradable material specially designed for packaging, is suitable for use in one-way boxes or blister packs. The starch-based material has promising properties, it is transparent and flexible, and in water or wet soil micro-organisms can degrade it within days to carbon dioxide and water.

Minor additives make the special starch, on which the material is based, easy to process. Among the applications foreseen are the envelopes around cigarette packets, insets in chocolate boxes, sanitary products, and gardening and agricultural materials and tools. The material may win a significant share of the protective packaging market, currently dominated by polystyrene.

A "second generation" biodegradable plastic which contained over 50 per cent cereal starch, petroleum based non-toxic synthetic polymers of molecular weight lower than polyethylene and hydrophiles has been developed. The time for biodegradation is from one to six months to a flexible film and several years for a hard plastic film.

Biodegradable plastic that consists entirely of starch and water, degrades in soil to form water and carbon dioxide.

Most existing "biodegradable" plastics contain up to 15 per cent of starch by weight, but they fail to decay completely because the remaining constituents resists degradation. Starch would normally disintegrate during processing unless blended with the non-degradable components.

In Japan researchers have developed a polysaccharide-based natural plastic. These plastics are processed from starch-like compound that exist naturally in materials such as comand crab shells and degrade in a month.

Researchers are developing biopolymers that degrade in soil in two to six weeks. Biopolymers can also be assimilated by the human body, opening the door, to new medical uses, including surgical sutures, glues, for broken bones and innovative drug delivery systems. The light-sensitivity and piezoelectric effects of these polymers may also make them useful as electronic materials.

The major application for biodegradable plastics at the moment is in carrier bags. Plastic containers will however, be the largest. Mulch film and garbage bags will also figure prominently.

BIODEGRADABLE PLASTICS — II

A plastic bottle discarded today will be around for between 200 and 400 years.

The problem is that overflowing landfills cannot continue to absorb the million tonnes of plastic trash created each year. In terms of volume, that's about 20 per cent of all trash. There are no simple solutions to reducing waste. The best solution, is a combination that includes biodegradable plastics, platics recycling, and composting.

The term "biodegradable plastics" means many things. Currently, "biodegradable" is used to describe both bio-disintegratable plastics that distintegrate into polymer chains, and biodegradable plastics that break down into water and carbon dioxide.

"Landfills are passive environments and most things shouldn't be in landfills". If paper doesn't degrade well in landfills, why should plastic?

"Everything eventually degrades", Nature-plants, animals, even humans are degradable. Dead organic materials are eaten by microbes and are converted to mulch. What we are trying to do is to modify synthetic cycles to be a part of the eco-system.

''Plastics are good materials but are bio-resistant''. In other words, the microbes that degrade organic materials like starch and paper can't easily degrade plastics made from petroleum. The reason lies in the way microbes metabolise their food, and in the molecular structure of the plastic. Conventional plastics are composed of many simple hydrocarbon molecules strung together in long chains called polymers. Plastics have a very long degradation time because microbes can only metabolise the polymer chains from the ends of the chain two molecules at a time.

To shorten the degradation time ''we must incorporate a biodegradable trait into these synthetic polymers so they go to mulch.'' The first generation of so-called biodegradable plastics were designed to do just that. Those plastics were made of (1) polymers, (2) about six per cent starch — usually cornstarch, which does biodegrade, and (3) a pro-oxidant which enhances the reaction of the polymer with oxygen in the air.

Like most materials, first generation biodegradable plastics don't degrade well in sanitary landfills primarily because of the lack of moisture. In a compost pile, however, results are slightly better. There microbes attack the starch, causing the plastic to disintegrate, and the pro-oxidant begins a chemical reaction that breaks the polymer chains into short links. When this plastic has distintegrated, it leaves behind in the environment short polymer chains that further metabolise at as yet undetermined rates.

From an industrial viewpoint, the problem with the pro-oxidant additive is that it begins to work as soon as it's added to the plastic. ''The solution is to encapsulate it into a biodegradable polymer matrix so it can't act until it's attacked by the microbes in the soil''.

Since these plastics were first introduced, scientists have made significant improvements. New materials are composed of more starch up to 97 per cent in some cases and, therefore, less plastic. Other, more revolutionary, findings have yielded plastics that degrade 100 per cent within six weeks after they are discarded. They are now available commercially.

One such plastic is made by micro-organisms. Broadly, this plastic is known As PHA (poly-hydroxyalkanoate) and is a member of the polyester family. A subset of the PHAs, called PHBs (polyhydroxy-butyrates) has received a great deal of attention because it was the first of this family to be commercialised, although there are four other members of the family, each with valuable properties as biodegradable plastics. Their production methods and applications are virtually identical to each other although their molecular composition and stiffness vary. This family is the only type of plastic made by microbes and is one of the few types that breakdown completely in the environment.

The relatively new process of growing the large quantities of PHA-producing microbes needed to produce usable amounts of the plastic is similar to brewing beer. The microbes produce the biopolymer PHA, and stockpile it internally as granules, upto about 95 per cent of their weight. During the lean times, the microbes eat the PHA granules. The process resembles a bear storing fat for hibernation.

That is what makes PHA plastics biodegradable—the PHA producing microbes also carry the chemistry for breaking down PHA. Of course, the more PHA is processed industrially, the greater the likelihood of introducing biodegradation-inhibiting characteristics into the plastic material. "The rule so far," is the organisms seem to break down PHA, but there are limitations. Inside a cell, it's biodegradable. The PHA is 100 per cent biodegradable rather than being merely biodisintegratable.

Commercially, the PHA-producing bacteria are grown in fermentation vats with molasses of methanol as food. The preferred method of producing stimulation of large-scale PHA production is to grow the microbes as large as possible and then to deprive them of an essential nutrient and gorge them with sugar so they begin forming PHA. Once the PHA is formed it is removed from the microbes, sometimes using a steam and caustic mixture to break them open. The PHA is then heated to form the biogradable plastic.

Changing the food, bacteria, or conditions in the vat can change the properties of the resulting plastic. These plastics have the same strength and flexibility as commonly used petroleum-based plastics and can be melted, moulded, and shaped as required. However, they should not be used in applications requiring resistance to high temperature or impact. Even the same production techniques used for petroleum-based plastics (injection moulding, extrusion blow moulding, melt casting, or spinning) can be used without retooling the manufacturing plant. The main differences are that PHA-based plastics will not float, and they will degrade in a wet, microberich environment in about six weeks, depending upon their thickness.

Biodegradation is an inevitable long-range solution.

As PHA becomes more common, existing sewage treatment plants could be used to degrade the plastic. It would also be possible to use the microbes that produce PHA to degrade PHA products and convert them back into original PHA that could then be processed into new plastic materials.

Another promising technology involves laminating paper cups with 100 per cent cellulose-starch-based materials.

Composting is the new word among plastics researchers and environmentalists alike, and not without reasons. It may be possible also to compose starch-based plastics.

''Composting is very straight-forward.'' About 20 per cent of what goes into a landfill is compostable material such as lawn clippings and leaves. The final product of composting is humus which can be used as a mulching material.

Biodegradable plastics are viable alternatives and are moving slowly into the marketplace. By 2000 biodegradable plastics are expected to account for only three per cent of all plastic packaging waste. Instead of waiting for biodegradable plastics to solve the problem of our rapidly filling landfills, citizens, industry, and science must all take advantage of the technologies that can be used today—recycling and composting—while working toward greater use of biodegradables.

5. Nuclear Waste

NUCLEAR WASTE MANAGEMENT

About 30 nations use nuclear energy to generate one-sixth of the world's electricity. Some 530 nuclear power reactors are now in operation, and another 96 are under construction, according to the International Atomic Energy Agency (IAEA).

Three kinds of waste

The IAEA recognises high, intermediate, and low-level forms of nuclear waste. The classification is based on several factors including the waste's source and temperature, and half-life. The greater the number of long-lived radioisotopes the waste contains the higher the level of waste (generally speaking, a long-lived isotope is one with a half-life of more than 30 years).

High-level waste is produced in two ways : by reprocessing spent nuclear fuel to recover isotopes that can be used again as fuel, and by merely using a reactor's fuel rods, which contain long-lived isotopes when spent. This waste is also at first very hot.

Intermediate-level waste consists of reactor by products and other materials, including equipment and tools that have become radioactive. It is less radioactive, and less hot than the high level type, but cannot be handled or transported without shielding.

Low-level waste, which can be handled without shielding, comes largely from commercial power, medical and other non-military sources, and consists, mainly of ordinary refuse (metal, glass, paper, rags and the like) that has been contaminated by radioactive materials. Though the least harmful, low-level waste is also by far the most voluminous.

Hot, high-level waste is created during the first few stages of reprocessing, when uranium and plutonium are separated from spent fuel, leaving behind dozens of long-lived radioisotopes in a solution of nitric acid. This soultion is then diverted to stainless steel tanks, which are equipped with cooling systems. The liquid may also be treated to neutralise it and precipitate the radioactive elements, forming a sludge.

POSSIBLE SOLUTIONS

Melting powdered glass with dried, concentrated high-level wastes, a procedure known as vitrification has been used on a large scale.

Although vitrification is being embraced by nuclear experts around the world, scientists are convinced they have a better way : sealing the waste within a form of synthetic rock, or ''synroc'', created from three titanate minerals and a little metal alloy. Several independent researchers have found synroc-clad waste superior to vitrified waste in important characteristics, including durability, resistance to leaching and irradiation by neutrons, and suitability to burial in deep boreholes.

Lacking a permanent respository, vitrified wastes are now being stored in air-cooled vaults containing compartments with multilayer tubes. Each consists of a plugged inner tube, which holds the waste, while cooling air is circulated in a surrounding outer tube. A vitrification plant about to begin operation will use a similar arrangement.

Some countries are operating centralized, temporary storage sites, called AFR, or away-from-reactor, for high-level wastes.

Designing a repository must take into account seismic, volcanic, hydrologic and geochemistry characteristics of the surrounding geologic formation to ensure that the wastes are isolated virtually indefinitely. The key technical issue is preventing, redioactive material from contaminating groundwater, which could poison food and drinking water, so water flux, velocity, and flow path at a prospective site are carefully studied in different geological formations. The repository must also be engineered to remove the heat generated by high-level waste. The possibility of human intrusion is another important risk factor.

A main counterargument — put forward by both environmentalists and some technical experts — is that there is no pressing need for a deep repository. The technology exists to store the wastes on the surface, this lets them cool significantly until they are eventually disposed of underground.

The capacity of existing storage sites can in fact, be expanded. Replacing the original aluminium storage racks that hold fuel assemblies in storage pools with new racks made of more expensive materials, such as borated steel, can triple capacity.

While high-level disposal problems will persist beyond the turn of the century, more progress has been made in getting rid of the low-level type. The volume of these wastes — more than 100 times that of high-level materials — can often be reduced by compaction shredding, and incineration. Because the radioactivity involved is small, low-level wastes rarely require radiation shielding.

On the intermediate level, reactor resins, solidified chemical sludges, or used equipment are radioactive enough to need shielding during handling. More problematic are mixed wastes, which contain both radioactive and other hazardous materials, such as organic chemicals, which makes safe disposal difficult.

NUCLEAR WASTE DISPOSAL — I

The disposal of industrial waste, some of which is highly toxic and dangerous to handle, has become a major problem. The traditional methods of disposal are in rivers, streams or the sea and burning and burial in landfill sites. These methods have been the same for decades, but there is increasing awareness of the environmental pollution they cause and greater care is being taken in discarding dangerous waste materials.

The nuclear power industry produces a good deal of waste, some of which is potentially dangerous because of its radioactivity. However, because of the extremely high energy and density of nuclear fuels, the quantity of waste produced per unit of electricity generated is considerably less than that of a comparable coal- and oil-fired power stations.

Furthermore, because the nuclear industry is a relatively young one, it has, from its inception, been subject to strict national and international regulations. They govern all stages of nuclear construction, operation, handling, storage and disposal of radioactive waste.

This is produced in varying quantities at virtually every stage of the nuclear fuel cycle, ranging from the mining and initial processing of the uranium ore and the use of nuclear fuel elements at the power station, to the reprocessing (cleansing) of the spent fuel when it is removed from the reactor. The use of radioisotopes in industry, agriculture, education and medicine also gives rise to substantial quantities of waste.

Low, intermediate and high

Nuclear wastes are categorised in three levels — low, intermediate and high — and they may emerge in gaseous, solid or liquid form. The category assigned to a particular waste is determined by the type of radiation emitted (alpha, beta or gamma) and the concentration of the radioactivity.

These two factors essentially define the potential of the material for causing harm to human and animal life as well as the environment.

Low-level waste comprises laboratory refuse such as protective clothing, rubber gloves, tissues, swabs and plastic bags, besides liquid effluents from sinks and drains and radioactive gases present in ventilation systems. Although very low in radioactivity, the waste is high in volume because of its bulky nature.

Inermediate-level waste consists mainly of cladding material removed from spent fuel elements and, to a lesser extent, of contaminated pipework, control rod mechanisms and structural materials found in the core of reactor.

High-level waste originates entirely from the reproccessing of spent fuel elements — a form of nitric acid containing fission products (the impurities) extracted from the components. It is sometimes described as heat-generating waste because of the warmth produced by the decay of its high, concentrated radioactivity.

Vitrification describes a process that enables high-level liquid waste to be converted into insoluble blocks of glass. At the same time, the process reduces the original volume of waste as much as two-thirds. Here, the waste conveyed from the tank storage area into the vitrification plant's own highly-active liquor tank. From here it is fed at a slow but continuous rate into what is known as the calciner stage.

This comprises a stainless steel tube which is inclined at a three per cent slope and rotated inside an electrically-heated furnace. As the liquid waste slowly trickles down the calciner it is evaporated, dried and partially de-nitrated to produce a fine, dry, brown-coloured powder known as calcine.

The calcine is fed into a vessel where it is mixed with crushed glass in the ratio of 15 per cent calcine to 75 per cent glass. The mixture is heated by an induction furnace to a temperature of 1150^{0}C at which the glass melts and the calcine dissolves to create a molten, vitrified mixture of glass and high-level waste.

Vitrified product store

When the mixture reaches a certain level in the melting vessel—

usually after about eight hours — it is poured into a stainless steel container measuring 1.3m in height and 420mm in diameter. The mass of liquid normally held by the product container is 400 kg; equivalent to 150 litres.

The product container is lowered on to turntable where it is allowed to cool. After being fitted with a lid secured by fusion welding it is inspected by remote-controlled surveillance equipment and transferred to the adjacent Vitrified product store (VPS).

The store consists of a series of rectangular compartments constructed on a base slab of reinforced concrete. Inside each compartment is a storage channel and thimble-shaped tube. The tube accommodate up to ten product storage cotnainers stacked one above the other.

The store has a capacity of 8000 containers with provision for increasing storage at a later stage. The vitrified blocks are cooled by a naturally convected airflow.

The containers can be kept in the VPS for at least 50 years before permanent disposal.

NUCLEAR WASTE DISPOSAL — II

Emission of particles from a radioactive source, and of its ability to penetrate and damage human (and animal and plant) tissue, is the crux of the problem. The property of radioactivity that gives rise to this kind of penetrating radiation derives from the spontaneous disintegration of atomic nuclei of certain heavy metals (radium, actinium, uranium, thorium). (The unit of radioactivity is the becquerel.) The atoms of these metals are inherently ustable and liable to transform into other element and its isotope (another form of the element differing from it in atomic weight and nuclear properties) this disintegration, called 'radioactive decay', can take different forms. Alpha, beta and gamma radiation may be emitted, each with its own characteristics. The ultimate end-product of radioactive disintegration is an isotope of lead.

Among the natural sources of radiation are the earth's soils and surface rocks: the atmosphere with its radioactive gases, radon and thoron, which seep into it from the earth's crust; and cosmic radiation which comes from outer spaceand the sun. The major part of the remaining one-tenth of the radiation from human activities comes from medical uses, non-nuclear industries, agriculture, and research. The nuclear industry contributes only a miniscule 0.1 per cent of the total radiation we receive. The difference in hazard from radiation from the nuclear industry and from other sources lies in the concentration and high-level from the one source and the dispersion and low-level from the other. Animal and plant life systems have adapted to natural radiation over millions of years, but are incapable of dealing with the new, lethal mancreated doses.

Radiation is always potentially harmful however small the dose. A rough straight-line relationship was used between radiation dose and probability of illness induction. With high-level radiation, these range from overburden of the body's repair mechanisms and consequent inability to cope, to stimulation of these mechanisms leading to survival of deranged cells and their transformation to malignancy. Thus, damaged body cells may die or become cancerous, causing, for example leukaemia (The unit of radioactive dose is the sievert). The tendency has been to progressively reduce it.

The effects of radiation are sufficiently alarming. Given that we cannot abolish nuclear power, what can be done to contain the radiation and the workers' (and others') exposure to it to the barest minimum? What specifically, are the sources of this radiation in the industry? Do they vary in their lethality? Let us turn our attention to the kinds of waste produced, which together make up the dreaded nuclear waste in the public mind.

Nuclear power waste embraces a big range of materials, from the first stages with the mining of uranium. It ranges from uranium mine tailings, to contaminated equipment and operators' clothes, to lethally toxic liquid high-level waste from reprocessing. Those in charge of the nuclear industry tend to bracket these with radioactive waste from hospitals, other industries, and agriculture. They point to coalgenerated power as responsible for millions of tons of carbon dioxide, sulphur dioxide, nitrous oxide and coal ash, the last with its many poisonous contents.

A large pressurised water reactor (PWR) produces irradiated fuel, by now 270,000 times more radioactive than the ore from which it was derived. Stored this is the only volume to be dealt with. If, however, this spent fuel is reprocessed, it produces further high-level medium-level and low-level waste.

High-level waste, is the most radioactive, and is in liquid form; medium-level waste is less radioactive and consists of many items like the cans which encased the spent nuclear fuel rods, contaminated equipment from the many stages of nuclear reaction and associated operations, and sludges from treatment processes; and low-level waste, of low radioactive content, comprises operators' protective clothing, equipment in fringe areas of operation, and other discarded material from these areas.

Concrete vaults now hold containers of compacted waste, and when full, will be covered with earth. Liquid waste is treated, physically and chemically, in water tanks, to reduce its radioactive content, and discharged into the environmentalists, however object to any dumping at sea.

Medium-level waste is encapsulated in cement in steel drums. The drums are then kept in specially built stores in the short term, for eventual disposal in underground caverns at selected sites. Underground burial has had its problems from leakage, resulting in contaminated groundwater.

High-level waste, from the reprocessing of spent nuclear fuel, has a 96 per cent uranium content which can be made into new fuel, about one per cent plutonium, leaving a remainder of 3 per cent liquid waste. Until recently, this liquid waste was concentrated by evaporation and stored in double-walled, stainless steel tanks within concrete walls. Now, it is being converted into solid blocks of a glass like substance (aborosilicate glass) by a process called vitrification, and a new plant to do this has been built. The liquid waste is turned into powder, dissolved in molten glass, and poured into stainless steel containers where it solidifies. The liquid is thus reduced to a third of its initial volume. The containers of vitrified waste are placed in water-cooled stores. Where they are allowed to cool over many years. They will eventually be moved to underground caverns for further cooling, a process taking some 50-70 years. It is thought that even more secure storage will be achieved by final burial in the seabed in appropriate corrosion-resisting containers.

The basic problem of nuclear waste is that it becomes harmless only after millions of years. Plutonium, for example, takes 25,000 years for only half of its radioactivity to decay to a safe level (known as its half-life), and can take hundreds of thousands of years more to decay altogether. Each nuclear reactor produces around 200 kg of plutonium each year, an amount that is enough to inflict cancer on every living human being. Present world plans allow 'secure' storage of high-level waste for the next thousand years. This must, inevitably, be largely speculative.

DISMANTILING NUCLEAR PLANTS

Dismantiling of nuclear plants will take longer than building them.

With over a score of nuclear power reactors in different countries now under various stages of decommissioning that can span over five decades, there is a proliferation of in-depth studies on decommissioning NPS, with respect to safety, modes of disassembly, waste disposal methods and, above all costs.

Given current levels of technological tooling and innovation, dismantling normal reactor poses little problems.

Decommissioning becomes problematic only in reactors that have been affected by accidents, such as those of Chernobyl in the USSR, and at the Three Mile Island (TMI) in the United States.

While at Chernobyl, the affected reactor has been completely entombed, at TMI, the authorities cleared up those parts of the NPS zone outside the 'Biological Shield' (comprising the concrete structure within which the reactor vessel is housed) recently.

For purposes of decommissioning, the plant is classified into three zones. Zone I refers to the inside 'concrete biological shield' of the reactor, Zone II to the plant and structures outside the 'biological shield' but containing radioactive material as in the pumps, pipework, control system parts, and fuel storage pool, and Zone III comprises the non-radioactive parts like the sea-water system, and transformers.

The decommissioning of a plant is itself phased out in three stages with varying time spans. In stage I of recommissioning, which could go upto five years, the fuel, in which the bulk of the radioactive material in any plant is contained, is taken out of the reactor core entirely, and sent away to the reprocessing plant.

Likewise, in the first stage, the coolant heavy water (in the case of heavy water reactors is also drained for purification and reuse. The inventory of the radioactive material in the station is thus drastically reduced; while Zone I is kept under surveillance, all accessible radioactive materials in and around the reactor building are removed and the non-radioactive parts in Zone III are also cleared.

In the second stage of decommissioning, which could run to another five years, the components and equipment in Zone II, such as team generators, pipework, valves and pumps, which are likely to have some radioactive deposits settled on them, are removed using well-tested chemical treatment means.

The stage is then set for the third and final phase of decommissioning, namely disassembling the reactor vessel itself. Most of the radioactivity in the reactor vessel is due to Cobalt-60, which has a half-life of about five years. Hence, most of the radioactivity in the reactor-vessel would come down a thousand to a million fold in 50 or 100 years respectively.

Thus, if the removal of the reactor vessel and concrete block around it is deferred for about 50 years, during which time there is very little possibility of this radioactive material appearing in the environment, it would be possible to take advantage of a substantial reduction in radiation levels.

Underwater plasma cutting devices could be used to break up the larger vessel into smaller parts to facilitate handling and disposal. The concrete wall of the biological shield could be cut through with 'water jets' and the water prevents any dust from getting air-borne.

The water used in the dismantling operation could be collected, purified and disposed off safely.

At present the Shippingport plant (pressurised water reactor of 90 mwe capacity) in the United States, is the first commercial-size nuclear power plant to be safely and cost-effectively dismantled completely down to the third stage. Another reactor in the US at Elk River site and one in Germany at Niederaichbach have reached the third stage of decommissioning.

In the category of less than 100 mwe capacity reactors, two French, two German and one American and Swedish reactors are in the first stage of decommissioning. In the second stage of decommissioning in this category are two reactors in the US, and one each in Switzerland and Germany.

NUCLEAR POWER REACTORS

A nuclear energy-producing reactor is an assembly of equipment in which a nuclear reaction can be sustained to provide large quantities of heat which can then be converted into other energy forms. It comprises primarily nuclear fuel, moderator, and control rods in steady operation. Electricity from a conventional reactor source is generated in much the same way as in a thermal (coal—or oil—fired) power station. Heat converts water into steam which drives turbo-generators (turbines driven by impact of jets of steam coupled to electricity generating machines) to produce electricity. (In a hydroelectric power station, water jets do the same job.) The only difference is the source of heat. With coal and oil the heat comes from burning the fuel. With nuclear power, it comes from splitting of atoms of uranium, a radioactive, hard, grey metal.

This splitting is called 'nuclear fission'. The nucleus (comprising positively charged protons and uncharged neutrons) of an uranium atom is split into two or more fragments plus one or more neutrons when hit by a travelling neutron. The reaction is initiated by absorption of the neutron, and releases energy in the form of heat. The excess neutrons produced go on to strike the nuclei of other uranium atoms and cause further fission, thus setting up a self-sustaining chain reaction. These two circumstances of an exceedingly large amount of energy release in the fission reaction and the production of excess neutrons to sustain a chain reaction make it possible to design nuclear reactors in which self-sustaining reactions occur with continuous release of energy.

Within the reactor, the fission occurs in a controlled way. The fuel is placed in metal tubes which are lowered into the moderator (water, heavy water or deuterium oxide, or graphite). As neutrons collide with the nuclei of the moderator, their energy and speed are reduced until a thermal equilibrium is achieved within the reactor structure. The rate of reaction is regulated by control rods of special alloy which have a high neutron capture rate (boron carbide), which absorb excess neutrons above those needed to maintain a constant reactivity level. The rods are inserted into the core and moved in and out as required to provide control over the degree of reactivity.

The reactions that take place in the reactor are complex, but this is a simple account of the process. The total mass of the fragments into which an uranium atom is split is less than the combined mass of the original nucleus and the bombarding neutron. The missing mass is transformed into energy or heat.

The heat thus produced is extracted by a fluid coolant and transferred to a boiler. From then on, a nuclear power station operates like any other thermal power station. The amount of heat produced is massive in relation to the quantity of fuel required within the reactor — fundamentally different from coal and oil. One tonne of uranium can produce as much energy as 20,000 tonnes of coal. Furthermore, uranium has no other significant use and is available in plenty in the world, while oil and gas reserves are estimated to last about 75 years and coal about 200 years. These, then are the basics of a nuclear power station.

Today, some 400 nuclear power stations produce about 20 per cent of the world's electricity. But the proportion is higher in some of the industrialised countries, notably France 75 per cent and Belgium 61 per cent. In Britain it is about 22 per cent. These countries are committed to nuclear power expansion. Others are uncertain about future plans or have even halted their programmes. The U.S., 19 per cent, is undecided, and Sweden, 45 per cent, has gone further and planned to phase out nuclear power stations altogether by the year 2010.

The fuel used in nuclear reactors is uranium. It can be reprocessed after use and recycled into more fuel. Reprocessing recovers a by-product, plutonium, produced from uranium by neutron absorption in the reactor, which can then be used as fuel in fast reacts, which themselves produce fresh fuel, thus increasing the energy obtainable from uranium about 60 times. Radioactivity is present at every stage of these processes. Almost everything that comes into contact with the reactor process, from clothes worn by the operators to the coolant, becomes contaminated. It is also in the nature of the process that all reactors leak a little and minor accidents are commonplace.

So, people working in nuclear power stations are constantly at risk. High doses of radiation are instantly fatal; low doses are cumulative and lead to illnesses, mainly cancer. Scientists within the nuclear industry set acceptable radiation levels for workers in it. But many outside believe that there are no such levels and no such thing as a "safe dose". Experience of a few years, for that is what we have had is clearly not enough to set safe levels if the health of people is the top priority.

Hundreds of types of nuclear reactors have been designed and tested, based on uranium and plutonium as fuel. The ones now commonly being built are the Advanced Gas-cooled Reactor (AGR), the Pressurised Water Reactor (PWR), and the new fast breeder reactor. None of them is 100 per cent safe, which is true of any technological system. Nuclear installations have been built, and are being built, to be as safe as possible in available circumstances. In the erstwhile Soviet Union self-imposed shortcuts which opted for flawed reactor design and overlooked stringent safety and operational protection for political reactions interfered with application of the best known technology and eventually led to disaster. Even in an advanced and industrial safety-conscious nation as the former West Germany, there were some 300 minor nuclear accidents in a single year, 1988. It is thought that the statistics are no better anywhere else, if not readily available. The risk of a major accident is now put at 70 per cent every six years, an alarming prospect.

So then, what do we do? Do away with nuclear power stations altogether and seek to develop alternative energy sources? Or, carry on with the nuclear programme, inbuilding maximum safety devices into the best reactor designs? There is an insistent cry from many environmentalists to scrap nuclear power and switch to natural energy sources—hydro, wind, solar, wave, tidal, geothermal—as safer and cleaner than nuclear, coal, oil, or gas. Of these, only hydro power has been harnessed on a big scale, causing huge environmental damage while conferring the benefits of power, irrigation, flood control, etc. Of the others, there is little sign that any of them is as yet capable of contributing substantially to the world's energy needs. Where local conditions have been favourable, some have produced useful energy. Even so a great deal of research and development still needs to be done to determine their reliability and environmental suitability.

In such a scenario, it is idle to condemn nuclear power outright. Increased industrialisation and growing populations induce huge demands for energy, as the developing countreis go for higher living standards. Fossil fuels — coal, gas and oil—will not last for ever. Are we prepared to do without the energy that is being and could be derived from nuclear power in the next half-century? The answer will be a thunderous 'no'. In the years to come, perhaps in the second quarter of the next century, the scientific dream of plentiful and pollution-free nuclear power from nuclear fusion (as distinct from nuclear fission) may come true.

NUCLEAR TESTS EDUCATION

The study on "Radioactive Heaven and Earth", predicts that about 430,000 people will die of cancer because of exposure in this century to radioactive fallout, and millions more will die of exposure in the centuries to come. The figure is based on United Nations estimates of the amount of radiation released from more than 500 atmospheric tests around the world and on a recent report by a committee of the National Academy of Sciences that sharply increased the estimate in the damage to human health caused by a given quantity of radiation. A moratorium on testing is favoured.

A need is felt for "re-education" of the public to accept testing "within a matter of 100 or so miles of their homes".

"Adequate education of the people must launched concerning the radiological hazards resulting from atomic explosions, should be realistic in nature, with the view to giving the public a correct understanding of this matter in order that the hysterical or alarmist complex now so prevalent may be corrected". Pronouncements that create "an unhealthy, dangerous and unjustified fear of atomic detonations" should be avoided.

The main risk is transferred to future generations. The testing has created and is continuing to create unstudied, uncharacterised, and unlicensed nuclear dumps for high-level radioactive wastes.

Plutonium is a toxic substance that remains dangerously radioactive for hundreds of thousands of years, as a major problem in the high-level nuclear waste tanks.

Safety problems in the storage of highly radioactive nuclear wastes left over from bomb production, are to be attended.

Long term radioactive waste disposal sites have to be found.

Large quantities of radioactive wastes are being explosively injected into fractured underground cavities without serious concern about future containment of the long-lived radioactive materials. Some of these materials have already spread faster than expected.

RADIOACTIVE WASTE DISPOSAL

Half a century after the world's nuclear industries began accumulating radioactive waste, not a single country has found a safe, permanent way to dispose of it.

Geologic burial—packing radioactive waste hundreds of metres deep in the earth's crust—is the safest long-term option for irradiated fuel rods and other "high-level" waste.

But there is the possibility of disturbance of the waste by groundwater movement, geological activity or human intervention. Brine seeps through the walls. This corrosive groundwater could eat away steel containers and create a radioactive slurry that wends its way into a nearby aquatic pond or river.

Volcanic activity could resume before stock-pile of radioactive waste cools down. To rely on a system of redundant engineering barriers is safer rather than simply deep burial.

As to the engineering challenges, it is now clear that geologic burial cannot guarantee that waste will remain sealed off from the biosphere. Geologic disposal is nothing more than a calculated risk.

Above-ground "temporary" storage appears to be the *de facto* "solution" facing many nations well into the 21 century. Temporary storage at least offers the best chance of buying time while the search for a technologically and socially feasible route continues.

NUCLEAR ACCIDENTS

The accident at the nuclear power station in St. Petersburg (former Soviet Union) takes into account the dangers of nuclear power. The existing nuclear reactors are like potential atom bombs.

Both the kind of technology used in the Russian nuclear reactors and the quality of construction and the presonnel maning them made them most vulnerable to future accidents.

While debating a moratorium on any future construction of nuclear reactors or to close down gradually the Chernobyl reactor, it was revealed that the sarcophagus, the concrete cover built over the fourth reactor at Chernobyl which exploded in April 1986, had developed cracks through which rain and snow seeped into the reactor. There had been another accident in reactor 2 at Chernobyl in November, 1991. These incidents should convince Governments about the dangers posed by nuclear power.

The problems faced by people had not ended with the Chernobyl accident but had just begun. Each passing day new evidence was being gathered indicating long-term health problems faced by the population living within a certain radius of Chernobyl. There had been a visible decline in the birth rate and increase in the death rate since the accident.

Is a total ban on nuclear power called for in the light of the latest accident in a nuclear reactor ? "How many more accidents will it take to realise the global threat to the environment that is posed by a continuation of nuclear power". There is a greater than one in four chance of a meltdown in the oldest VVER reactors, located mainly in Eastern and Central Europe, within the next five years. It is also said that the world's most dangerous ageing nuclear reactors were the 22 magnox reactors in the U.K.

6. Global Climate

MONITORING GLOBAL CLIMATE CHANGE

The phenomenon of global change in the atmosphere, in the oceans and on land using a wide range of instrumentation is being studied. The sophisticated remote sensing and acoustic techniques employed are producing precise and comprehensive measurements that provide keener insight, into understanding the Earth's climate.

Researchers have been taking careful measurements since the 1987 discovery that the ozone shield over the South Polar regions had declined by 50 per cent. Chlorofluorocarbons (refrigerants and aerosol propellants) have been blamed for producing the decline in the stratospheric ozone. And the chlorine dioxide recorded in the atmosphere over the Antarctic proves that CFCs are indeed responsible, as its molecule shows what amounts to a chlorine atom from a CFC that has broken apart a molecule of ozone.

Several instruments, such as satellite-borne TOMS, the Total Ozone Mapping Spectrometer and SBVU, the Solar Backscatter Ultraviolet Spectrometer, have been used to measure the widening ozone hole over the Antarctic. These work by detecting the absorption of sunlight as it scatters or reflects from the Earth and is absorbed by ozone at its characteristic ultraviolet wavelengths.

STUDYING CLIMATE AND THE OCEANS

Computer models of the atmosphere can forecast the weather a few days ahead. On the time scales of climate the behaviour of the atmosphere is regulated by exchanges of heat especially, with the land, the ice caps and, most importantly, the ocean. A useful climate system model must take note of these exchanges.

Weather forecasting uses the most powerful computers and is supported by a global network of thousands of observing stations reporting every few hours. Coupled atmosphere/ocean climate model will require even more computer power and a system for observing and reporting the behaviour of the ocean. How will these needs be met?

Fortunately, climate predictions are not needed on the exacting timetable of weather forecasts and the power of computers is increasing rapidly. Already it is possible to do a great deal by allowing big computers longer to do their work.

The problem of data from the ocean is more difficult. Comprehensive descriptions of the ocean, both in present state and its variability, and of its exchange with the atmosphere are needed but they are not available as yet.

International Collaboration

Oceanographers and meteorologists have long recognised this problem and the need to harness international effort to solve it.

Oceanographic study undertaken will involve scientists from more than 30 countries. Its goals will be to (i) develop and collect data to test, (ii) make models of the ocean/atmosphere system, and (iii) find ways of predicting the effects of climate change as time goes on.

The first of these goals will be achieved through three core projects during the Nineties.

Core Project-1—The Global Description—will describe the circulation and chemistry of the ocean. Much of the work will be done as a single survey—the WOCE hydrographic programme—to which many Nations will contribute.

Core Project-2—The Southern Ocean—will study the way that the main oceans are linked by the Antarctic circumpolar current and the formation of cold dense bottom and intermediate water.

Core Project-3—The Gyre Dynamics Experiment—in the north Altantic will identify the processes responsible for basing-wide circulation.

Operational Phase

The second goal looks at longer time-scales and will assess how representative the WOCE data sets are of the long term behaviour of the ocean.

Observations from satellites can tell much about what is going on at the surface of the ocean—winds, waves, surface elevations, temperature—and, indirectly, about what is happening beneath. They will help reach the second goal. These will be complemented by measurements from an already partly established network of ships, buoys and submerged instruments.

GLOBAL WARMING

Nitrogen oxide fumes emitted by aircraft exert 30 times as great an effect on climate as the same fumes emitted at ground level by industrial processes and the burning of fossil fuels like coal and oil, British scientists have calculated.

The nitrogen oxides react with other chemicals in the air to create ozone in the troposphere, or lower atmosphere. At higher altitudes, in the stratosphere, ozone blocks ultra-violet rays from the sun that can cause cancer in humans. In the troposphere, ozone is toxic to plats and animals, and it is one of a number of tropospheric gases—carbon dioxide is the chief one—that trap heat. Many scientists believe that increase in the gases as a result of human activity will cause the earth's climate to warm substantially in the coming decades.

If the growth in aircraft traffic in the 1990's continues at current rates, nitrogen oxides would result in the trapping of enough solar energy to raise the earth's temperature by about two-hundredths of a degree Fahrenheit.

The heating effect of ozone, carbon dioxide and all the other heat-trapping gases may be diminished or enhanced by a number of competing factors.

SOIL CARBON IN GLOBAL WARMING

World's soil contains some 1,500 gigatonnes of carbon, held in organic matter. Large amounts of it, if released as carbon dioxide (CO_2) into the atmosphere through global warming, could add to the greenhouse effect and raise the global temperature.

Soil organic matter levels will fall because of greenhouse effect, atmospheric CO_2 will increase and so will the temperature.

How far this destabilising feedback might be counteracted by other factors to reach a new equilibrium?

A multi-compartmental model predicts carbon turnover and its consequences. It also takes soil moisture into account.

The amount of organic matter in the soil depends upon the input amounts of plant and animal debris and the rate at which the organic matter decays. Decomposition is strongly related to temperature, so when the soil warms up, the rate of decay rises and more CO_2 is produced.

The soil normally contained 31.6 tonnes of carbon per hectare when under wheat crops, with an input of 1.5 tonnes of carbon per hectare each year. If the temperature rises by 1.5°C (all other factors remaining constant) the soil's organic matter falls to a new equilibrium of 26.0 tonnes per hectare, and still lower to 17.5 tonnes if the temperature goes up by 5°C.

A projection of global warming predicts the temperature rise if—

(a) annual production of other greenhouse gases such as oxides of nitrogen and CFCs stays as it is now.

(b) industrial release of CO_2 continues at five billion tonnes of carbon annually and

(c) production of plants is not affected by changes in climate and CO_2 levels.

F. N. : The model has been tested against known data on the decomposition of organic materials, obtained over periods ranging from 13 months to nearly seven years.

7. Natural Disasters

FIGHTING NATURAL DISASTERS AND CALAMITIES

A natural disaster is unforeseen, severe and immediate. Pollution, ozone depletion in the stratosphere and global warming came in this category. Natural disasters include cyclones, earthquakes, floods, drought, heat and cold waves, landslides, avalanches, flash floods, severe thunderstorms, hail, low level wind shears, and microburst and disasters like forest fires, gas leaks etc.

The destructive potential of any natural hazard is estimated basically by its spatial extent and severity. Spatial extent upto which the effect of a disastrous event could be felt may easily be classified into small, medium and large scales. The phenomenon extending from a few kilometres to a few tens of kilometres are termed as small scale. However, low level wind shear and microburst, which are extremely hazardous for aviation, are of very short duration and localised phenomenon, could be categorised as microscale.

Growing industrialisation and unjustified exploitation of natural resources have brought our eco-system to a verge of non-reversibility and imbalance. This has led to athreat from a set of natural hazards like pollution, global warming and ozone depletion on large or global scale. The medium scale natural hazards such as cyclones, earthquakes, floods and drought are the familiar and frequently occurring devastating phenomena.

The effects, warning techniques, and some aspects of mitigation of these natural hazards are discussed below:

The management aspects of disaster may be classified as (a) early warning system, (b) rescue operations, (c) relief operations, (d) rehabilitation, and (e) long range planning. The most important is the early warning system. Unless sufficient advance notice is available, evacuation of the population likely to be affected cannot be undertaken.

There are two aspects of early warning system. One is the availability of an effective technique to forecast the disaster with its extent and the other is effective communication of the same to the authority responsible for rescue operations.

In some phenomena, like cyclones, floods, etc., the time available to respond to the hazard is of the order of a few days. Hence early warning, communication, and rescue operations are possible. But, in a few cases like flash floods, microburst, etc., the response time is of the order of few minutes only, which calls for a very fast early warning and efficient communication system. The human-activity-induced hazards like pollution and global warming have already started showing their precursors, giving sufficient time to control and avoid these hazards by long term planning. On the contrary, in earthquakes no proven method has yet been evolved to give any prior warning and so post hazard mitigation is the only alternative.

Critical role

The role of communication in disaster mitigation is extremely critical. The various types available for disseminating disaster warning as well as arranging mitigation are (a) land line links, (b) underground cable links, (c) wireless links, (d) microwave (LOS) and (e) satellite link.

The only effective communication which is likely to remain completely or partially unaffected is the satellite link. This assumes that the earth stations at the two ends are suitably located to remain unaffected. Further link between the earth station and the affected area is usually through microwave/land line, which has its limitations, as it may breakdown.

The most effective way of warning is DWS (Disaster Warning System) for issue of cyclone bulletins. This could be extended to the entire earthquake/floodprone areas.

For effective two way communication, VHF/UHF links should be established from every earth station to the affected zone. Use of existing VHF/UHF link can be made. The linking of these with VHF/UHF stations would be a cost effective and reliable communication system for disaster warning and mitigation.

The status of prediction models for some calamities as well as the communication system are discussed below :

Cyclones

Tropical cyclones, the msot destructive of nature's phenomena, are known to form over all tropical oceans except over South Atlantic and the South Pacific, east of about 140°W. An intense low pressure area in the atmosphere is formed before after the monsoon. It is associated with fierce wind and heavy rainfall. Horizontally it extends from 500 to 1000 km and vertically from surface to about 14 km.

Severe tropical cyclones cause considerable damage to property and agricultural crop. The Tropical cyclones are tracked with the help of (i) regular observations from weather network of surface and upper air observation stations, (ii) ships report, (iii) cyclone detection radars, (iv) satellites, and (v) reports from commercial aircraft.

Warnings are issued by the area cyclene warning centres.

A system known as Disaster Warning System (DWS) transmits cyclone warning bulletins through INSAT-DWS to the recipients. This consists of the following elements :

(i) The cyclone warning centre for originating the area code of the districts and disaster warning message.

(ii) The earth station located near the cyclone warning centre with uplink facility in C-band and suitable communication links.

(iii) The C/S band transponder on board INSAT.

(iv) The INSAT-DWS receivers located in cyclone prone areas.

Earthquakes

Various models using the statistical methods have been evolved which are used to predict the likely intensity of quakes in a seismic zone. Still the exact time cannot be predicted.

The epicentral parameters of earthquakes in a region are determined through a network of seismological stations.

Earthquake recurrence of different magnitudes is studied either by using the Gutenberg frequency magnitude relationship or through the use of Gumbel's statistics.

For a safe and reliable communication in the quake prone areas, satellite based communication is essential. DoT channels usually break down due to fissures/cracks in roads/fields etc. It will be useful to have battery backup to the transreceivers in various districts as power supply is the first casualty in an earthquake.

The basic information required for flood forecasting is rainfall data of the catchment area of the river. With sophisticated high powered S-band radars with DVIP facility, it is now possible to estimate the rainfall upto 200 km around the radar site. This system is used extensively in the U.S. for estimating the rainfall potential in the catchment areas of major rivers and issue of flood forecast.

The use of radar for the precipitation estimate is based on the principle that the amount of echo return from a volume of cloud depends on the number and size of hydrometers in it. The empirical relationship between the echo return and the rainfall rates has been developed for various types of rain.

Using fast switching digital circuits, the return video is digitised, integrated, normalised and contoured into standard six or seven rainfall rates. Observations taken every ten minutes can be cumulatively added and averaged to give 24-hour rainfall over the region. Through appropriate modes, the information from a number of radar sites can be sent to a central office where powerful computers process the data and yield the overall rainfall potential of the weather system.

The advantage of using radars for hydrological work lies in the fact that information over the inaccessible region is available without actual human intervention. Of course, there are many assumptions which do not always hold good, thereby introducing large errors in the result. But with suitable calibration with actual gauge measurements, correction factors could be applied. Another advantage of the radar measurement is that it cuts the time for the collection of rainfall data, thereby increasing the lead time available for rescue/evacuation efforts in the region likely to be affected.

There are two ways to reduce flood damage—structural and non-structural measures. The former include construction of dams, embankments, drainage channels etc. This has not helped much as population has moved into areas where flooding used to occur and has been controlled due to structure. Whenever flooding level is higher than what the structure can hold occur the result is devastating.

The non-structural approach calls for removing populations from the flood plains. Another important aspect is to reduce the silting of rivers. Afforestation in the catchment areas, along the river banks helps in breaking away of soil and would help in maintaining the effective river volume.

DISASTER MANAGEMENT
(Coping With Natural Disasters)

Despite the tremendous advance man has made in the field of science and technology, mankind remains virtually helpless in the face of natural disasters such as floods, cyclones, avalanches, volcanic eruptions and earthquakes which causes widespread loss to life and property. The most recent earthquake in Uttar Pradesh which took a heavy toll of human life and property is a classic instance of this most unfortunate situation.

According to a recent study on an average 1241 lives are lost in India every year on account of floods. Nearly 26.5 million people are affected, and over 75,000 houses are destroyed by floods and cyclones alone. The total loss of property is estimated at Rs. 800 crores annually.

A review for the last few years suffices it to show the diversity, frequency and severe consequences of natural disasters—cyclones, floods, tornadoes, avalanches and earthquakes. Every time a tragedy occurs, there is a talk of advance steps and better and enthusiasm die as soon as the calamity stops. Disaster management is therefore, what is most called for.

Disaster management is not only search, rescue, relief, first aid, cure of ailment, rehabilitation and reconstruction, but also the education of both the volunteers and the people in the disaster prone belt, in the form of "dos and don'ts" besides appreciation of the nature of the calamity.

Safety rules

(1) When an area is covered by a cyclone warning, the residents should continue normal activities, but stay turned in to radio or television for further news, keep alert and ignore rumours.

(2) The residents should plan their time before the storm arrives and avoid last-minute hurry which might leave the area marooned.

(3) People should keep calm until the emergency has ended. The ability to meet an emergency will inspire and help others. People should leave low-lying areas that may be swept by high tides or storm waves.

(4) Enough drinking water should be stored in clean bathtubs, jugs, bottles, and cooking utensil; town water supply may be contaminated by flooding or damaged by hurricane floods.

(5) Battery-powered equipment should be checked and kept ready for use. Radio may be the only link with the world outside. Emergency cooking facilities, lights, and flashlights will be essential if utilities are interrupted.

(6) Vehicles should be kept fuelled. The petrol service station may

be inoperable for several days after the storm strikes, due to flooding or interrupted electrical power.

(7) People should stay indoors during a hurricane. Travel is not advisable when winds and tides whip through an area. At least one door should be kept open on the leeway side of the house.

Tornadoes spawned by cyclones are among the worst killers. A tornado warning means that tornado has actually been sighted. When an area receives a tornado warning, people should seek inside shelter immediately, preferably below ground level. If a house has no basement the inmates should take shelter under heavy furniture on the ground floor or in a small room that is away from outside walls and windows. If a tornado catches a pedestrian outside, he/she should move away from its path at a right angle. If there is no time to escape, lying flat in the nearest depression such as a ditch or ravine is very safe.

When Earthquake strikes (Rehabilitation Measures)

When an earthquake strikes an area for a minute or two the "solid" earth moves like the deck of a ship. What one does during and immediately after the tremor may make life-and-death differences for the entire family, and neighbours. The following rules will help :

(1) The motion is frightening but, unless it shakes something down on top of the residents it is harmless. The earth does not yawn open, gulp a neighbourhood, and slam shut.

(2) If it catches one indoors, it is better to stay indoors. Taking cover under a desk, table bench or in doorways, halls, and against inside walls will help. People should stay away from glass.

(3) Use of candles, matches or other open flames, either during or after a tremor should be avoided.

(4) While in the open one should move away from buildings and utility wires to avoid danger from falling debris and live wires.

(5) If one is in a moving car, one should stop as quickly as safety permits, but stay in the vehicle.

People living in a flood disaster prone area should take to tree-planting. Trees help check soil erosion and direct impact of flood water on buildings.

If called upon to evacuate, people should not wait until the last moment, but accept lcoal administration advice and proceed to safer areas.

People should not try to cross a stream or pool of water unless they are certain about the depth of water and the condition of the route. Water may hide a washed away bridge or cut out road. While crossing water drive in lower gear so as to avoid splashing of water. Remember, as the brakes may fail after moving through water they should be tried as soon as the vehicle is out of water.

Filth left behind may cause outbreak of epidemic. Inoculation and avoiding use of contaminated food or water etc., is advised.

A State and district level disaster preparedness plan may be evolved for every State and district in disaster vulnerable areas.

State Governments and Central Government and voluntary agencies may create groups of various levels—village, district, state and national —which could be activated immediately a disaster strikes to coordinate the efforts coming from all sources.

The voluntary agencies coordinate their work after a disaster. A system of effective and continuing consultation may be established among voluntary agencies, between government and voluntary agencies and among various agencies involved in the same locality.

Involvement of volunteers from within the community for counter-disaster activities should be encouraged to stimulate participation of village people who take interest in protecting their own welfare and working towards their own development.

The United Nations, including its specialized agencies, the League of Red Cross societies and other voluntary agencies should render more financial and technical assistance to the regions concerned. Appropriate international organisations should widen the operation of satellites and extend their coverage.

By adopting various scientific methods mentioned above through an integrated approach and through a concerted effort, any nation can certainly cope with natural disasters.

F.N. : The recent devastating Earthquake in Maharashtra and the adjoining States has taken a toll of 10,000 lives, ruined 30,000 families, flattened the whole villages and collapsed the houses and the catastrophe is the worst tragedy and horrifying, turning residences into huge graves, and trapping people under the debris, too difficult to extricate either living or dead. Relief measures included deploying people to unearth the bodies under the debris, treating those alive with medicines — anti-trauma and anti-bleeding drugs, reconstruction and rehabilitation.

MONITORING CYCLONES FOR EARLY WARNING

Tropical cyclones are one of the most violent manifestations of nature and potentially the deadliest of all meteorological phenomena. It is a unique combination of gales, heavy rain and mountainous waves. When the wind speed in a rotating tropical disturbance exceeds 64 kmph it is termed a cyclone. However, the tropical cyclone becomes dangerous when the wind speed exceeds 120 kmph. At this stage the system is termed as severe cyclone storm with a core of hurricane wind.

About 80 storms occur around the world a year. The average annual damage has been estimated at about $ 1 billion and the annual death toll about 15,000 over the last 30 years.

Destructive winds and torrential rain from the cloud mass both lasting several hours are common associates of tropical cyclones. But in severe cyclones the result of the wind stress on sea dwarfs these two in disaster potential on the coast.

Surge : Cyclones are associated with high pressure gradient and consequent strong winds. These in turn generate storm surge. Storm surge is an abnormal rise of sea level near the coast with devastating effect. In some parts of the world the tides in a cyclonic storm has many times exceeded 12m, about height of a four-storey building.

The height of storm surge depends on the intensity of the cyclone and the topography of sea bed near the point of landfall. (Sea level also rises due to astronomical high tide.) The size of the tidal waves increases when peak surge occurs at the time of high tide. Vulnerability to storm surge is not uniform along Indian coasts.

Wind : The winds in a severe cyclone of hurricane intensity may reach 150-200 kmph or more. Such violent winds can blow off kutcha structures like huts with thatched roofs, tiled and asbestos roofs and cause damage to telecommunication and power supply installations. The flying debris can be lethal. The gale can wreak havoc on crops in the field, orchards and plantations. Such winds can uproot trees and disrupt road communications.

Rainfall : Heavy rain is always associated with tropical cyclones. Floods in the plain and landslip in the hills are the major destructive effects of intense rain. In cyclone, damage potential of the wind is increased because of the accompanying rainfall. Cyclone detection radars having a detection range of about 400 km, Satellite picture receiving

(APT) equipment are available in the cyclone warning centres to receive satellite cloud imagery from the polar orbiting satellite. Cyclone warnings are the very basic information for cyclone distress prevention and mitigation.

These are disseminated under two stage warning scheme whenever any coastal belt is expected to experience heavy rain, gales, tidal waves in association with cyclonic storms. The first stage known as the 'cyclone alert' is issued 48 hours in advance of the expected commencement of the adverse stage weather over the coastal area. The second stage warning known as the 'cyclone warning' is issued 24 hours in advance. Both the cyclone alert and warning messages are passed for repeated broadcast. Warning messages are also issued to port, fisheries, railways.

INSAT based Disaster Warning System (DWS) is being operated. Under this scheme the cyclone warning messages in the regional language are sent directly to the public through INSAT. About 100 DWS receivers have been installed in some of the coastal districts. Detailed plans for distress mitigation are reviewed every year before the cyclone season. While the loss of life due to cyclone has decreased over the years in almost all countries due to precautionary measures, the loss of property has been increasing steadily due to rapid economic development in these parts.

The cyclone is tracked by the INSAT and radars track the system as it came within their range.

'Cyclone alert' message followed by bulletins is transmitted till the storm crosses coast, and weather improves.

With these timely warnings and adequate preventive measures, the suffering can be mitigated.

The tropical cyclone is not without its goods effects. Many areas get much rainfall from this cyclone. A single cyclone in fact fills up the reservoirs which have been parched for months by drought.

EARTHQUAKES

An earthquake begins deep inside the earth as an expression of its internal energy, shifting masses of rock and mineral inside that set up the waves we feel as tremors. There are a mission quakes in one year on the earth, most of them on the ocean floor. The disastrous ones, causing large loss of life are thankfully under a dozen a year. This focus of an earthquake is inside the earth and the area on the earth's surface directly above the focus is its epicentre. The deeper the focus the less the damage on the surface, and the more shallow the earthquake is the more destruction it causes. The most severe quakes release a total energy approximately equal to a hundred large atom bombs. They can make the earth vibrate enough for it to be felt across the globe.

Earthquake waves are generally either surface waves that move over the crust or bodily waves that move through the interior of the earth. They are studied by an instrument, seismograph, which is a ray of light drawing the depth of the tremors on sensitized paper. The severity of an earthquake is measured on a scale developed by Richter. On the Richter scale, each unit means a release of energy ten times greater than the previous one. So a quake measuring 7 will be ten times stronger than 6 and goes up to 9.

Our planet is withstanding a lot of pressures. Not only the internal or geological ones, but also planetary gravitational pulls. In addition to this are the pressures that mankind and our technology imposes on the earth while harnessing its natural resources. The large volume of water collected in reservoir of big hydro electric projects exerts a tremendous pressure on the earth. This has been known to cause occasional dam-related earthquakes.

As part of our respect for environment conditions it is therefore important that each development project should be thoroughly examined for its effect on ecology. To undertake development without regard to the environment can endanger many innocent human lives.

Those of us who have experienced an earthquake know that there is a rumbling sound, doors, and windows rattle, beds and tables in the room shake, and if we are standing we feel the floor moving away from under our feet. If the quake is major, then the earth cracks open and either water gushes out or lava comes out to form volcanoes. This is known as an earthquake i.e. when the earth rumbles and trembles and finally with one big lurch opens up to form huge openings called ''fissures''.

Under the hard earth crust, there are large thick, rocky plates floating on molten rock. As these rock plates, nearly 140 miles thick, are not on solid ground they constantly move or rub against each other causing earth tremors. In a day there are about 2000 such movements, but most of them are so mild that we hardly notice them and they pass without causing any harm.

The intensity is measured by the instrument is so sensitive that it can record even the slightest earth tremor.

Most earthquakes occur along the sides of the plates, especially where they interact—spreading, parts sliding or underthrusting each other. Scientists have been able to spot quake-prone areas. It is interesting to know that the Indian sub-continent is slowly moving northwards and pushing up the Himalayan mountains.

Though scientists have been able to locate sensitive areas, they have not been able to say when it would strike. They are now observing very carefully the slight tilt in the land, changes in the well water levels, release of Radon gas and other changes in nature which will enable them to foretell an earthquake.

MAIN CAUSES OF EARTHQUAKES

1. Earthquakes are caused by stresses within the Earth. These stresses are due to certain unstable subterranean conditions in the geological formations brought about by the presence of fault planes and other stressful environments like the internal heat of the earth. Such instability can lead to the shifting of rock bodies within the Earth, thereby creating shock waves, which, under suitable conditions, can result in an earthquake. Minor slippages of rock bodies may result in mild tremors that occur ever so frequently in many parts of the world. (Assam experiences such tremors very often.)

2. Earth is an elastic medium possessing certain well-defined physical properties. It is not, however, a homogeneous medium. It is a layered medium with several geological formations deposited one over the other over millions of years of geological time. These formations are made up of marine and continental sedimentary rocks as well as igneous rocks formed by many volcanic eruptions.

3. The various geological formations of the Earth can propagate the shock waves both through the body of the Earth as well as over the surface. The shock waves, which cause earthquakes, fall under two broad categories; surface waves and body waves travel through its body, penetrating the Earth, getting reflected and refracted in the process when they encounter different geological formations. The body waves can be further classified as primary or P waves and secondary or S waves. An earthquake creates both surface and body waves, but it is the P and S body waves that reach seismograph station earlier than the slower surface waves. Earth being a non-homogeneous and layered

medium, the velocities of propagation of these waves differ from one medium to the other.

4. While the P waves can travel through a solid as well as liquid medium, S waves can travel only through a solid medium, and are stopped by a fluid (liquid or gas) medium. It is this difference in their propagation characteristics that led to investigations which established the fact that the Earth possessed a molten, liquid core extending over 3,500 kilometres from the centre. The propagation velocities of P and S waves are also different in a given medium. P waves are faster and making use of this difference in the velocities, scientists are able to locate the epicentre whenever an earthquake occurs.

5. The instrument used for detecting earthquakes, is the seismograph; the first of its kind was invented towards the end of the 19th century by John Milne, a British scientist. It was a simple device consisting of a block suspended by a spring inside another block attached to the ground.

6. It was the American seismologist, Charles Richter, who, in 1935, invented a scale for measuring the intensity of an earthquake. Known as the Richter scale, it ranges from 0 to 9 on a logarthmic scale. An intensity of two on the Richter scale is roughly 30 times — and not just twice — the intensity of one. The highest known intensity so far recorded on this scale measured about 8.5. This was notched up by an earthquake that hit Alaska in North America in 1964.

7. Earthquakes of mild or moderate intensity occur frequently over many parts of the world, but generally cause minimum damage to life and property, and pass of as tremors. (But a few of them, like the one that hit Latur and the one that devastated Los Angeles and Jakarta Islands recently turn out to be giant killers.)

8. When a major earthquake occurs it sends shock waves over a large area and tremors of varying intensity can be felt at places far removed from the actual scene of disaster.

9. An earthquake occurring below an ocean bed generates gigantic swells in the ocean. As long as these swells remain over the deeper parts of the ocean they are long and gentle waves. But when they make landfall, the shelving coastline causes them to rise and crash one over the other successively unleashing in the process, towering piles of water several metres high. If the inhabitants of the coastal area are not forewarned and evacuated in time, loss of human lives can reach calamitous proportions.

10. Not all regions of the world are earthquake prone. However, some regions that were once believed to be outside the danger zone, after all may not be quite so. (For example, the Deccan Plateau, once believed to be outside the seismic belt, may not be a safe zone, as brought home by the Latur disaster. Earth scientists will, sooner or later, develop the capability to accurately predict earth-quakes, but, cannot avoid its occurrence.)

PREDICTING EARTHQUAKES FROM SPACE

Accurate prediction of earthquakes and volcanic eruptions could become a reality, due to new kind of satellite 'remote sensing'.

The images are taken using a sophisticated technique called synthetic aperture radar (SAR). The technique is so sensitive that ground movements of as little as 34 mm can be detected, even without obvious rupture at the surface.

Measurement of ground deformation following earthquakes is usually achieved by careful monitoring of the movements of well-placed 'ground stations', sensitive equipment stationed in areas where earthquakes are thought likely to happen. Even when the positions of these ground stations are 'fixed' very accurately by satellite ranging, each one has to be installed and maintained by human observers — so there's practical upper limit to what can be achieved by this means, usually a few tens or hundreds of points scattered over the ground.

Detailed knowledge of topography before a major earthquake strikes depends on having enough of the right instruments deployed in the right places at the right times. This information is vital, because ground deformation reveals the changing pattern of stresses in faults under the ground, and may yield information on where and when the next earthquake will strike.

Satellite-borne SAR takes a picture every time it passes over a potentially earthquake-prone area, storing it for future use.

Potentially life-threatening earthquakes and volcanic eruptions are often presaged by deformations in the Earth's surface. Satellite SAR pictures could be used to learn more about how and why these deformations happen, whether the deformations that precede earthquakes or eruptions are characteristic (and thus diagnostic) and—in time—be used to deliver accurate earthquake or eruption warnings.

EARTH'S HOT CORE

Why hasn't the core of the earth cooled off?

The earth is cooling at present, but, very slowly.

The thermal history of the earth is principally a question of the balance between heat sources and heat sinks. When the sources are greater than the loss, the heat of the earth increases, and vice versa. Over the last four billion years, the average temperature has fallen by a few hundred degrees centigrade at most, perhaps as little as 200 degrees Fahrenheit.

Throughout the 4.5 billion-year history of the earth, the heat balance has changed. Initially, in the first 500 million years or so, there was a large increase.

MAGNETIC SUBSTORMS

What triggers a magnetic substorm in which there is a sudden major disturbance in the Earth's magnetic field?

By calculating the spiral paths taken by individual electrically charged particles the complex interaction between the Earth's own magnetic field and the solar wind is found out.

In the search to identify the circumstances that trigger a magnetic substorm, the characteristic spiralling motions of particles in the magnetotail on the nightside of the Earth, where its magnetic field is drawn out into an extended region have been recorded.

If the Earth were totally isolated, its magnetic field would closely resemble that of a familiar bar magnet—a "dipole" field—but the effect of the solar wind of electrically charged particles streaming past causes the field to become greatly distended on the side of the Earth away from the Sun.

The magnetotail acts as a kind of storehouse for the energy brought into the Earth's magnetic field by the solar wind particles. The build-up to a magnetic substorm starts with an increase in the rate at which energetic particles arriving from the solar wind are fed into the Earth's magnetosphere. The magnetotail becomes even more distended until something "gives" and the field springs back to resemble more closely an undistorted dipole field this is the so-called substorm. In the process, energetic particles are fed in closer to the Earth producing effects such as the enhanced aurora.

The motion of charged particles in the magnetotail has two distinct components—spiralling motion around the magnetic field lines and a bouncing motion across the current sheet between the regions where the magnetic field is in opposite directions. When the frequency of either motion is much greater than the other, the particles travel in orderly spirals. If the two frequencies are similar, chaotic motion sets in. The disruption caused by the onset of chaos may be enough to provoke the massive reorganisation of the magnetic field observed as a substorm.

This motion develops in a magnetic field changing over time and the motion to become chaotic, a substorm start. The data collected about the charged particles in the magnetotail should reveal any chaotic motion. The time at which it occurs can then be compared with when substorms are recorded by regular magnetic field measurements on the Earth.

ANTARCTIC ICE BUILDUP

Warming of the cliamte is causing a "significant increase" in the accumulation of ice in East Antarctica, rather than a depletion as widely predicted, American and Japanese scientists have found.

This is based on dating layers in sea-floor sediment extracted, where tongues of continental ice sometimes push far out to sea before breaking off.

A similar finding has been reported by Australian scientists who have measured snow accumulation 60 to 180 miles inland. They have found that accumulation has been increasing and recently reached 20 per cent above the mean.

The researchers of Estimate that the accumulation should contribute to a slight to lowering of global sea level, contrary to the trend reported elsewhere. Snow accumulation close to the coast is, however, affected by oceanic weather and many not show what is happening to the continent as a whole.

8. Cleaner Technologies

CLEANER TECHNOLOGIES AND PRODUCTION FOR POLLUTION CONTROL

"Prevention is better than cure". The conventional approach to pollution prevention at the end of the production pipeline through effluent dilution or treatment is now obsolete as well as inadequate and expensive.

It is better to adopt cleaner technologies that produce less pollutants rather than try to clean up the mess made by polluting technology. This message is particularly important for the developing countries for whom investment in pollution abatement systems is often an unaffordable proposition.

Cleaner production is not just a matter of process technology but an attitude to production that takes into account environmental management. Quite often just better housekeeping like stopping leaks and keeping the control instruments working properly can effect a substantial reduction of the effluent load by reducing wastages.

Recycling of products: The trend in the developed countries in environment management is to reduce the volume of pollutants and increase the use and recycling of waste products. Reduction of pollutant level is achieved by changes in process technology. Examples include use of membrane process for caustic soda production (instead of the mercury cell process which releases harmful mercury into the environment), the low pressure process of making polyethylene which halves the pollutant level compared to the conventional high pressure process, the use of water soluble resins instead of cyanide solutions in electro-plating, using hot fluidised-bed of alumina instead of molten cyanide salts for heat-treatment of steels, the use of salt-free solutions for dyeing cotton cloth, recovering and recycling caustic soda in the anodising of aluminium and so on.

There are a number of such improvements now available and can be accessed by anyone interested by approaching the International Cleaner Production Information Cleaninghouse (ICPIC) of the United

Nations Environment Programme in Paris and the Pollution Prevention Information Cleaninghouse (PPIC) initiated by the U.S. Environment Protection Agency (EPA). For industrial licensing authorities and pollution control boards as well as environmental activists in the Third World who want to check up on technologies likely to be imported into their countries, the ICPIC and PPIC are excellent sources of information.

Another new trend in the field of environment management is the "Cradle to Grave" analysis of a product which makes possible the likely environment impact of a product throughout its life-time and even after it has been discarded. Using such analysis, it is now possible to redesign products in such a manner as to reduce or minimise damage to the environment from its manufacture, use and subsequent disposal.

Yet another trend worthy of emulation by the developing countries with a fast developing industrial base like India is the close cooperation between industries and regulatory authorities on environment management.

FUEL CELLS FOR CLEAN POWER

A fuel cell is a device that converts the energy of a fuel directly to electricity and heat, without combustion. Because there is no combustion, fuel cells give off few emissions; and because there are no moving parts, fuel cells are quiet. Although a fuel is consumed, fuel cell plant is about 50 per cent more efficient than conventional gaspowered turbine power generators.

Several types of fuel cells are under development. Each uses a different type of electrolyte, an electrically conductive chemical, to react with the fuel. Phosphoric acid fuel cells are the most advanced and have been operating in power plants.

Molten carbonate is of simpler design, higher efficiency and lower material cost. In its normal state the carbonate is a solid salt, when heated to 1,200 degrees, it enters a liquid state. Natural gas — mostly methane — is first combined with 1,200-degree steam and then passed through a reformer, which converts the mixture to hydrogen and carbon dioxide.

The hydrogen is pumped into the anode, where it reacts with the molten carbonate, also at 1,200 degrees, producing water, carbon dioxide and electrons. Each hydrogen atom gives up two electrons.

The cathode grabs the two electrons, replenishing the electrolyte with carbonate ions. The movement of the electrons from the anode to the cathode generates the power. In effect, the carbonate ions migrate across the electrolyte and carry the charge with them, just as in a battery. Like a battery, the fuel cell produces direct current, which must be processed by an inverter to become the alternating current supplied by utilities.

Because the electrolyte is continually replenished, it does not need to be replaced. The process produces more water than it consumes, so the waste water can be fed back to become steam.

MATERIALS TECHNOLOGY IN FUEL CELL DEVELOPMENT

The direct conversion of chemical energy in a fuel cell has been one of the most challenging technological problems since a simple hydrogen/oxygen fuel cells was first demonstrated 150 years ago by Groves. Aqueous electrolyte fuel cells have been successfully used in space to power electrical equipment in the Apollo space craft and in the space shuttle but terrestial applications have been principally limited to a few specialised military and telecommunication applications.

The various types of fuel cells have common features. The characteristics of the interfacial three phase boundary play a major role in the electrochemical performance of a fuel cell, particularly in those systems incorporating liquid electrolytes. In such fuel cells, the reactant gases diffuse through a thin electrolyte film that wets portions of the porous electrode structure and react electro-chemically on the electrode surface.

Many Individual Cells

A fuel cell stack usually consists of a number of individual cells connected in electrical series and whilst a number of different stack configurations can be derived, in practice most fuel stack systems have a plenary arrangement. The individual cells are connected by a ribbed bipolar plate which provides a low electronic resistance between adjacent porous anode and cathode structures and also functions as a gas barrier between the fuel and oxidants streams in adjacent cells. The bipolar must therefore be a good electronic conductor and impermeable to gases. The ribbed channels provide a more uniform gas distribution and help to preserve the mechanical integrity of the various cell components.

The fuel and oxidant streams flow perpendicular to each other (crossflow) and the current flux is perpendicular to the gas flow. In addition to the fuel stack, materials technology also has an important role in such ancillary equipments as gas manifolds, gas supply lines heat exchangers, pumps, fuel processing plants, etc.

1. Phosphoric acid fuel cell (PAFC): Air/H_3 PO_4 $(H^+)_{aq}$ /H_2 (from hydrocarbons)

This system was originally developed for operation on hydrocarbon fuels. As the fuel cells can only utilise hydrogen derived from primary hydrocarbon fuels it is necessary for the system to incorporate a fuel processing unit which functions with the addition of steam as a reformer and shift convertor. The addition of this unit reduces the overall electrical/fuel conversion ratio to around 40 per cent which will undoubtedly limit market penetration to selected terrestial civil applications particularly those associated with combined heat and power (CHP) and for stationary dispersed power plants.

PAFC power plants are available in 40 KW, 200 KW and 11 MW sizes. Increasing concern over environmental pollution will undoubtedly favour the introduction of these systems in the Nineties.

Major breakthrough

A major breakthrough in PAFC technology occurred about 20 years ago with the development of appropriate carbon blacks and graphite cell components. The porous electrodes are now fabricated from a specially developed finely divided carbon black onto which crystallites (2.5nm) of the electrocatalyst (pt) are deposited. The carbon black particles (surface area 100-200 m^2 [g]) are held together with a PTFE binder (about 30-50 wt per cent) to form a porous component which is supported on a porous carbon paper substrate. This composite structure forms a stable three phase interface in the fuel cell, with H_3PO_4 electrolyte on one side (electrocatalyst side) and the reactant gas environment on the other side of the carbon paper.

These electrode structures perform satisfactorily for around 40,000 hours under current operating conditions and most development work is concerned with controlling the growth of the Pt crystallites at loadings of 0.25 mg/cm^2 and/or replacing the Pt with more active binary (e.g. PT_3V) and ternary (e.g. Pt-V-Co) electrocatalytically active Engel-Brewer phases. As the operating temperature and pressures of PAFCs increase, the corrosion stability of carbon containing cell components may become a problem and so alternative electrocatalyst supports such as TiC are currently under examination.

A stable porous separator material which incorporates the phosphoric acid electrolyte has also been successfully developed. This component is a PTFE bonded SiC composite material which also performs successfully under current operating conditions.

2. Molten carbonate fuel cell (MCFC)

Air/K_2CO_3–Li_2Co_3 ($CO^2{}_3$) Hydrocarbon.

The MCFC is one of the two high temperature systems which are regarded as "second generation" fuel cells. In principle high temperature operation can allow "in-situ" reforming of the hydrocarbon fuel and electricity/fuel conversion efficiencies about 60 per cent are forecast. The MCFC has been under development, and units in the range 1-10 KW are currently being evaluated.

3. Solid oxide fuel cell (SOFC) : Air (Y)O_2 (O^2) Hydrocarbon

This system incorporate a ceramic solid oxide electrolyte, with a typical composition of $Zr_{0.9}Y_{0.1}O_{1.95}$. The anion vacancies created by the introduction of aliovalent^{3+} ions on the Zr^{4+} sites produce an oxygen ion conductor which has a conductivity comparable to liquid electrolytes in the temperature range 900-1000°C. Because the electrolyte is solid the electrolyte management problems associated with other fuel cells types are eliminated. Moreover the high temperature operation allows *in situ* reforming of hydrocarbon fuels and so electricity/fuel conversion efficiencies should approach 60% (cf. MCFC system).

Attempts were made to construct SOFC systems and although small units (—10W) were technically successful and operated for more than 30,000 hours it was difficult to assemble strong tough ceramic components economically. However there is a resurgence of interest in SOFC systems due to recent developments in ceramic technology which allow the fabrication of monolithic ceramic electrolyte configurations incorporating thin strong tough components with high area/volume ratios. A tubular configuration was earlier used in various projects, the design has been scaled up to produce 5 KW.

Bright future

It appears that the future of ceramic SOFC systems is assured. Units up to about 100 KW should begin to be available for evaluation as CHP systems, and large multi MW units for central electrical power generation should begin to be installed shortly.

There is also much scope for retro-fitting ceramic SOFC systems to gas turbines to improve the overall efficiency for electrical generation. Power densities as high as 1 MW/M^3 have been forecast for SOFC systems and if these values are realised then these systems should also have a significant contribution to transport particularly as power units for ships, helicopters and heavy lorries.

4. Alkaline (bacon) fuel cell (AFC) : O_2/KOH (OH) $_{aq}H_2$

The special features of the AFC system including readily available fuel and oxidant already present for rocket motors made this system an attractive power unit for the space programme. 1.5 KW units were incorporated on the Apollo space craft used to take astronauts to the moon, and three 12 KW units provide the electrical supply for the space shuttle.

This fuel cell is now widely available for a variety of aerospace and defence applications as a reliable well proven engineered system. It is however very expensive to operate requiring pure hydrogen and oxygen and relatively costly electrode materials (for application in city buses, etc.).

5. Solid polymer electrolyte fuel cell (SPFC)

These fuel cells in incorporating polymeric membrane electrolytes were originally investigated for the space programme but were not used due to the superior performance of the AFC system. However recent developments in membrane technology allow much high power densities, and large development programmes are now under way to evaluate this system for a variety of defence (e.g. submarine propulsion) and space applications. It is, however, an expensive system requiring relatively pure O_2 and H_2 gases, and high loadings of noble metal electrocatalysts.

Materials technology has been principally responsible for the steady improvement in the performance of the relatively mature PAFC and AFC systems. A major contribution has been the fabrication of stable porous electrodes with an optimised interfacial three phase boundary. Indeed the preparation and distribution of the very small platinum crystallites provides an excellent example of nanotechnology engineering.

The combination of high efficiencies, low environmental pollution, quiet operation, and modular construction, all indicate that fuel cell system is a technology whose time has now arrived when we acutely need it.

FUEL CELLS ADVANTAGES

Fuel cells which use 'reverse electrolysis' to generate electricity can supplement other power sources.

However, fuel cells for power generation are not yet a commercial proposition. At present, the capital cost of these plants is rather high, but it is expected to come down with improvements in technology and larger volumes of production.

Fuel cells have several advantages. They are modular in construction and the modules could be added as and when demand increased. Fuel cells of a small capacity could be economical unlike as other generation systems which needed to be of a certain size (and hence expensive) to be economical. Fuel cells have high conversion efficiencies. They are silent, non-polluting, have no rotating parts and are highly reliable. They could use gas, oil or coal as fuel.

Hence, establishments like industrial units and hospitals in remote areas could use such fuel cells for on-site power generation. They could be an attractive option in places where a gas pipeline was passing nearby or coal was available cheaply.

With conventional electrolysis, water could be split using electricity to produce hydrogen and oxygen. Fuel cells run on the reverse principle—combining hydrogen and oxygen to give water and electricity, used electrodes of carbon bonded with polytetrafluoroethylene and phosphoric acid as electrolyte in its fuel cells. Steam at 180-200^{0}C is also produced and could be utilised for various purposes.

Oxygen is got from the atmosphere and hydrogen from natural gas, oil, methane, naphtha or coal (following the gasification route).

METHANOL-POWERED CARS

Cars designed to run on a fuel other than gasoline e.g. methanol are being experimented.

The main aim of developing cars powered by alternate fuels is not only gasoline conservation but meeting air pollution standards. Here, methanol has an advantage over gasoline. It burns more cleanly than gasoline, releasing less of the smog-causing hydrocarbons and nitrogen oxides.

Methanol has other advantages also. It has an octane rating of 100, compared with 93 to 97 for gasoline, that allows engines to run at higher compression and therefore more efficiently. Then, it has a better flame speed than gasoline, which speeds burning in the cylinders for a quicker pick up. Methanol also has another advantage in its high heat of evaporation which helps pull heat away from the engine. This would make it possible to reduce in weight of methanol-powered cars by using air-cooling radiator systems instead of the heavier water-cooling systems in gasoline-run cars.

Nevertheless methanol-powered cars have still some problems. For instance, burning methanol releases about half the energy that burning gasoline does. As a result, cars can go only half as far on a tank of methanol as they can on a tank of gas.

The consumers will also not appreciate the higher price tag on methanol vehicles, which are likely to cost more than the conventional vehicles.

The early models will be flexible fuel vehicles (FFVs) designed to run on different ratios of methanol and gas. These cars will be equipped with sensors that assess incoming fuel mixtures and then adjust fuel injection, spark plugs and emissions controls accordingly.

However, they are not as efficient as cars that use a fixed mixture of methanol and unleaded gas.

The flexible fuel vehicles have abolut 50 per cent efficiency advantage over gasoline-powered cars. A vehicle designed to run exclusively on methanol could probably achieve 30 per cent better efficiency than an equivalent gasoline model.

Another advantage would be the potential of reducing overall emissions of hydrocarbons and other pollutants by 90 per cent. With an ever growing awareness of the problems of global warming and air pollution, this would be a major promoter for these cars.

Methanol is easier to deliver than any other alternate fuel, such as natural gas. The methanol transportation infrastructure is almost half in cost compared to what other fuels would need.

HYDROGEN AS FUEL

Efforts are being made the world over to tap non-conventional energy sources such as solar, ocean thermal, wind, waves, currents, tides, thermonuclear, geothermal, etc. Energy from these sources has all the qualities of petroleum and natural gas. Some are only intermittently available, others are only available away from the consumption centres, and none can be used as a fuel for transportation.

The non-renewable, fossil fuels themselves cause some other problems. Their combustion products pollute the environment and produce acid rains. The main combustion product of fossil fuels, carbon dioxide, causes the "greenhouse effect". This leads to high temperatures, droughts, melting of the ice caps, floods and adverse climatic changes. Therefore, it becomes necessary to find an intermediary or synthetic form of energy which can be produced using the non conventional primary energy sources.

A hydrogen energy system could form the link between the new energy sources and the user.

Hydrogen is found in natural gas in small percentages. It forms 0.2 per cent of the atmosphere. There are very small quantities compared to the fuel needs of the world. Therefore, hydrogen must be manufactured using some primary energy source if it is to meet man's fuel needs.

Hydrogen as a fuel eliminates nearly all drawbacks of fossil fuels. It is clean, its residue (exhaust/ash) after burning is water or water by-product. It can be produced in large quantities.

Hydrogen has use for automotive purposes. Since it is the lightest of all fuels (nearly one-third the equivalent energy/weight as diesel oil or jet fuel) and is virtually totally pollution-free, it has a bright future in the aerospace industry. Buses and cars are already being run on hydrogen in many developed countries on an experimental basis, and the gas is now being commercialised for use in towns with heavy traffic.

HYDROGEN-POWERED ELECTRIC CAR

Electric car, powered by a hydrogen fuel cell rather than a battery, can compete on speed, acceleration and in economic terms with today's petrol-driven models.

Pollution-free Cars

Cars powered by hydrogen are pollution-free, producing only water as a by-product. Prototype hydrogen-powered cars have been around for at least two decades.

Plastic membrane

The fuel cell converts hydrogen into electricity. The cell is made of a fluorinated plastic membrane that is sandwiched between electrodes which are themselves sandwiched between two gas chambers. This arrangement works in a similar way to a battery, with the membrane — which is acidic when immersed in water — acting as the electrolyte.

The electrodes are porous to gases. Molecules of hydrogen pass from one gas chamber through the positive electrode to the membrane. Here, electrons are liberated and they pass into a circuit creating the current which drives the motor. The protons left behind are attracted to the negative electrode, where they recombine with electrons returning from the motor. The hydrogen atoms produced then react with oxygen from the other chamber to produce water. By stacking these cells end-to-end, current is produced.

Fuel cells using similar processes have been generating electricity in Apollo space missions. By reversing the process feeding electricity into the cell — the cell can split water into hydrogen and oxygen. This allows to recharge the car's hydrogen stock evernight using just water and off-peak electricity.

Other features

A tank full of granulated iron/titanium hydride stores enough hydrogen for nearly 500 kilometres of travel. But to force hydrogen to bond with this material needs high pressure. When the cell is being recharged from the mains, it converts water into hydrogen and oxygen.

These gases are produced in much greater volumes than the original water so the required pressure of 3.5 megapascals builds up as the cell is charged. When the car is running, heat is needed to free the hydrogen from the iron/titanium hydride in the tank. Waste heat produced by the cell heats the tank and frees the hydrogen.

HYDROGEN AS FUEL FOR ROAD VEHICLES

A number of alternatives to petrofuels have been studied, such as alcohols, methane, synfuel, solar photovoltaics, new generation storage batteries, fuel cells and, of course, hydrogen.

Of all these, hydrogen appears to be the most favoured alternative due to its high specific energy per unit weight, its almost universal availability as a component of water, good combustion characteristics and the fact that it is environment-friendly. The primary combustion products is water vapour and apart from low nitric exide fractions, there are virtually no harmful exhaust gases, in particular no carbon monoxide, hydrocarbons and particulates which are the bane of petrofuel combustion.

Because of its very low density, hydrogen has to be either compressed or liquefied or converted into a metal hydride before being carried as fuel on board a road vehicle. Compressed gas vessels and liquid hydrogen cryo-tanks were rejected, the first due to safety considerations (high pressure) and the second due to the high energy consumption for liquefaction.

That left hydride-forming alloys which enable a low operating pressure of hydrogen. The principle here is that certain metallic alloys absorb hydrogen in an exothermic reaction when exposed to the gas and release this absorbed gas when heat is applied to the metal hydrides, for use in vehicles, the hydride fuel tank consists of a closely packed bundle of alloy tubes over which hydrogen gas is passed at a service station. Cooling water is used during the exothermic absorption reaction and hot exhaust gas for the endothermic gas release reaction.

Some alloys have a high hydrogen storage capacity but can be technically used only on being subjected to temperatures above 300°C which is difficult with automobile exhaust gas. There are alloys which offer high storage capacities and release the gas at temperatures lower than 200°C (magnesium-copper) but these tend to dissociate during cyclic charge-discharge operations.

For vehicle application, a hydride meterial must permit a cold start and be suitable for heat exchange with normal coolants. The following parameters have to be optimised; storage capacity; hydrogen exchange kinetics; sensitivity to contaminant cases; cyclisation stability; activation behaviour; and reasonably priced industrial production.

An alloy consisting of titanium, zircon, vanadium, iron, chromium and manganese has been selected. The addition of copper or aluminium powder amounting to five to ten per cent by weight in relation to the hydride weight helped in stabilising the hydride powder, reducing fine dust displacement which can cause tube distortion and explosions. The pyrophorous (high reactivity with air), tendency of the hydride was also brought down.

The pressure/temperature relation of this heterogenous compound system (gas in equilibrium with a solid phase) is quite complex and very different from pure gas systems. A slight increase in temperature causes a very large increase in pressure. Therefore, as an added safety measure, the hydride storage units were equipped with safety valves which respond at 55 bars and vent the excess hydrogen to the atmosphere.

Problems

In designing the engine system to use hydrogen as fuel, the low ignition energy of the gas in air, the high lean burn ability of a hydrogen/air mixture, the low octane number of hydrogen and low density of the gas pose certain problems. These are:

(a) Irregular combustion cycles due to backfiring in the intake phase, pre-ignition in the compression phase, knocking during combustion and water accumulation at the spark plugs during cold start.

(b) Lower power density due to loss of volumetric efficiency.

There are two types of fuel/air mixture formation. In external mixture formation, the hydrogen is blown into the intake manifold. The advantage is that it is technically simple and requires only a low hydrogen pressure. The disadvantages are the low engine power and the marked tendency towards irregular combustion cycles.

With internal mixture formation, the hydrogen is injected directly into the combustion chamber during combustion, as in the diesel engine. Such a system prevents backfiring into the suction manifold and self-ignition in the compression phase. It does not exhibit knocking combustion because of the controlled injection of the fuel and has no losses in volumetric efficiency. Engine operation is characterised by a high power density with low fuel consumption. The limitations are the necessity of a high hydrogen supply pressure, complicated mixture formation system and problems which may occur with mixture homogenising at higher engine speeds.

HYDROGEN AS FUEL — PROBLEMS AND PROSPECTS

The use of the hydride storage technology as hydrogen storage system for the fleet vehicles and the low hydrogen supply pressure of this storage necessitated an engine with external mixture formation. Water injection was incorporated to cool critical components and residual gas so that the tendency to backfire and formation of nitrogen oxide could be reduced.

Problems relating to backfiring and cold start were fully taken care of by a newly designed high-voltage capacitor ignition system with electronic digital timing control, shielded ignition cables and a special spark plug. This ignition system met the following requirements:

High ignition voltage at full load.

High shunt insusceptibility during cold start.

Fast dropping of the residual energy in the ignition cable.

No induction in an ignition cable caused by adjacent cables.

The engine used was based on the standard 2.3 litre, 4 cylinder unit and had a compression ratio of 9:1. Compared to a standard gasoline engine of similar size, an approximate power reduction of 25 per cent was expected. Since a gas engine, in contrast to an engine operated with a leaded fuel, does not deposit a damping lead oxide layer in the area of the valves, suitable wear-resistant materials had to be used to combat elevated wear of valve seat rings and valves.

Various Problems

Operation of spark engines with hydrogen poses various problems with regard to lubrication caused by water forming as the main reaction product and injection of cooling water in the intake system. This results in substantial water loading of the engine oil. This not only increases susceptibility to corrosion but the use of conventional engine oils could lead to a stable oil-water emulsion with no lubricating effect which is even liable to freeze at low ambient temperatures. Two types of special lubricants were developed to tackle this problem — a demulsifying (hydrophobic) engine oil and a synthetic engine oil which forms a solution with water. Both types were used in the fleet test vehicles.

Besides using 100 per cent hydrogen as tuel, a mixed hydrogen gasoline fuel was also tried on the road. This combines the advantages of gasoline operation (large range, high energy density) with those of hydrogen operation (reduction of harmful emissions and improvement of specific fuel consumption). This concept foresees 100 per cent hydrogen operation during idling and 100 per cent gasoline operation at full load. In between, hydrogen and gasoline are injected according to the load requirement in a ratio allowing optimum consumption and exhaust gas values while assuring good driveability.

The electromagnetically actuated gasoline and hydrogen injection valves are operated by an electronic control unit in which the injection times for the two gases corresponding to load and speed are stored for the entire engine operating range. An electronically controlled accelerator system is used to regulate the engine by transmitting the load requirement signal to the mixture formation control unit and the microprocessor controlled ignition system.

Special spark plugs

The engine used for this mixed fuel concept was a standard 2.8 litre, 6 cyliner spark engine with a compression ratio of 9 : 1. In order to prevent problems during cold start as a result of condensed steam at the spark plug insulator, special spark plugs with a condensation water displacement space behind the insulator had to be developed. Throttling is only implemented during idling to reduce exhaust gas and consumption values even more.

Modifications were made to accommodate the hydride storage units, heat transfer systems and certain other changes to improve the braking behaviour and attain driving behaviour similar to that of standard gasoline vehicles.

Safety measures were incorporated for hydrogen storage (including operating pressure and temperature), engine fuels (hydrogen supply pressure, injection water pressure), engine operation (backfiring, position of hydrogen nozzles and control piston), hydrogen supply system (leaktightness) etc. Optical displays and acoustic alarms were added to inform the driver of possible malfunctions. In a dangerous situation, the monitoring system actively intervenes and shuts off the hydrogen carrying line by closing the solenoid valves.

In addition, measures were taken to ensure safe handling of the vehicles, such as:

— delayed start of ignition and fuel supply when starting the engine.

— inability to start when the refuelling tube is connected.

— interruption of hydrogen supply in the case of accidents.

— ventilation of storage and engine compartments to avoid hydrogen accumulation in the case of possible leaks.

Results obtained from the road tests revealed :

(a) *Operating experience* : On the whole this was positive with reports of good vehicle manoeuvrability and reliability in everyday traffic. Malfunctions were primarily caused by insufficient purity of hydrogen at the service station, inadequate cleanliness and demineralisation of engine injection water and external influences such as mechanical damage to the heat exchangers and exhaust gas systems due to inadequate road clearance.

(b) *Exhaust gases* : Hydrocarbon and carbon monoxide emissions for a fully hydrogen powered vehicle were lower than that of a gasoline vehicle with a three-way catalytic converter while the nitrogen oxides emissions were about the same. But with mixed hydrogen/gasoline operation, the emission valves were higher than in a gasoline fuelled vehicle with controlled three-way catalytic converter.

(c) *Refuelling duration* : In order to obtain almost 100 per cent filling of the storage unit, a refuelling time of 10 to 15 minutes for cars and 15 to 20 minutes for vans was observed.

(d) *Reactivation* : After about 140 to 150 refuelling cycles, the hydride storage unit has to be removed for reactivation by heat treatment. The original storage capacity was available again after reactivation.

Important conclusions drawn about hydrogen fuelled road vehicles

1. It has been shown that conventional vehicle concepts can be adapted to suit hydrogen as fuel and all types of combustion engines can be fitted for hydrogen operation. The component which suffers the most damage is the fuel tank and the key problem is the limitations imposed by the weight- and volume-related energy density. So, vehicle manufactures and operators need not change their whole way of thinking for hydrogen technology.

2. As of now, due to limits imposed by the almost total absence of a well-developed hydrogen supply infrastructure and the fundamental disadvantages exhibited by hydrogen drives with regard to range, payload, refuelling time and cost, hydrogen energy may be uncompetitive for road transport compared to conventional hydrocarbon fuels. However, the cost/benefit ratio may change in the medium term when fossil fuels get scarce, especially in countries with surplus hydel-power resources which can be used to manufacture hydrogen cheaply.

In the long term countries with favourable geo-climatic conditions of high year-round solar insulation can find it more beneficial to produce hydrogen through large solar energy plants than importing hydrocarbon fuels.

3. It may become necessary to introduce hydrogen energy to tackle the environmental problem of greenhouse and aciduous gases produced by burning hydrocarbon fuels. But even today, there is scope for taking to hydrogen fuelled vehicles in environmentally vulnerable areas such as industrial plants, shopping areas, underground mining, congested urban localities, hospitals, wildlife parks and nature preserves.

4. However, even if hydrogen drives in motor vehicles will attain significance a lot still needs to be done before this comes to pass, particularly in the sectors of production, storage and drives.

Present scene in India

Some work is being carried out in leading research institutions *e.g.*, developing a low cost, energy-efficient water electrolysis plant, investigation of semiconductor materials for photoanodes to be used in photoelectrolysis of water, liquefaction of hydrogen, investigation of metal alloys and inter-metallic compounds as hydrogen storage materials, use of hydrogen as substitute for diesel in direct injection internal combustion engines, and as substitute for gasoline in spark-ignited engines.

ENERGY SCENARIO, ITS MANAGEMENT, CONSERVATION IN PROCESS AND OTHER INDUSTRIES

I. Energy Scenario

Power generation in the country at the end of the century is expected to be 1.30 lakhs MW.

The Eighth Five-Year Plan started with a capacity generation of 70,000 MW and by the end of the plan period (1997) it could touch one lakh MW. Another 30,000 MW would be generated in the remaining three years.

The demand supply gap was 20 per cent and is expected to be the same even after additional power generation is made available by the end of 1997. The gap could be bridged to an extent by improving the plant load factor (PLF) of thermal stations and reducing the line losses. At present the PLF was 60 per cent all over the country and the power stations should improve it by at least two per cent each year.

An estimated Rs. 500,000 crores is required to meet the country's full power demand till the turn of the century. The power shortage had risen from 9 per cent in 1992-93 to 11 per cent in 1993-94 with peak shortages increasing from 20 to 25 per cent.

The average plant load factor (PLF) for the country has reached 61 per cent, a target envisaged for the end of the Eighth Five Year Plan and a considerable improvement over 57 per cent achieved in the previous year. Power generation also touched an all-time high of 323.32 billion units as compared to the target of 316.70 billion units for the year. Creation of additional capacity also exceeded the target of 4439 MW to reach 4538 MW. The requirement of additional power generating capacity has been pruned from the original estimate of 48,000 MW for the Eighth Plan to 30,500 MW, and is now going to be only 20,000 MW. As a result, the country will be short of 25,000 MW of generating capacity by the end of the plan period.

From 1980-81 onwards, the increase in thermal and nuclear power generation by far outpassed that in the hydroelectric sector. From the figures available, it is seen that addition to hydroelectric power during 1980-81 to 1992-93 was only 50 per cent; at the same time more than 300 per cent additional capacity was added to the thermal and nuclear power sectors. In spite of this quantum leap, the generation and distribution still falls short of the demand for power.

One reason for this shortfall is the idle capacity of the generator sector. The distribution network, too, has not kept pace with the increasing generating capacity. The result is that though generation has increased manifold, the distribution network has not improved. The reason is not far to seek: the SEBs are in the red and do not find funds to streamline the distribution network.

Inefficiency, strong worker's union, higher administrative costs with lack of realistic pricing policy, and hence, losses accrue. Much power has been given at nominal rates or free to many selected sectors.

With the private sector — both national and international entering the power generation field, assurance must be there of sufficient return on investment. There are tax holidays, assurance of payment of dues on time from SEBs. But coupled to this, there should be a realistic pricing policy where the investor does get a suitable return on investment.

Towards Cleaner and Efficient Coal-Fired Power Generation

Integrated coal gasification-based combined cycle (IGCC) is considered to be the future technology to address the shortcomings of the present-day coal based power plants. Besides higher efficiency and cleaner environmental performance, IGCC has a number of other advantages such as:

Smaller quantity of solid waste.
Low water consumption and
Capability for phased construction.

Combined cycle concept in IGCC involves conversion of coal to fuel gas in a gasifier under pressure and subsequent separation of the gas from particulate and chemical compounds harmful to the gas turbine. A steam cycle is thermodynamically coupled to the heat available exhaust gases of the gas turbine. (GT). This combination of GT and steam turbine (ST), offers significant increase in plant efficiency — over 40 per cent. Also, some of the industrial gas turbines can burn coal-gas, having heating value as low as 900 kcal/kg. with the necessary bleed air offtake for gasification. Today IGCC plant in single train, can be designed with a 200 MW GT coupled to 140 MW steam cycle.

The inlet temperature of a modern gas turbine ranges from 1000 deg. C to 1200 deg. C and the corresponding exhaust gas temperature from 470 deg. C to 600 deg. C. This facilitates the generation of high quality steam. In certain gasifiers the latent heat of the gas is also used to produce steam. In certain gasifiers the latent heat of the gas is also used to produce steam. This amounts to approximately 20 per cent of the steam turbine power.

Fuel Gas Processor

A coal gasification system comprises a coal feeder reactor and an ash removal system. In coal feeding system, coal is sized to the requirement and then fed into the gasifier either in dry form or as a mixture of coal and water slurry. In addition to coal, reactants like air or oxygen and stem are also fed. The proportion of these are determined by the gasification process. Devolatisation, partial oxidation, hydro-gasification, water-gas and other reactions with coal take place in the reactor at a process-specific temperature. The product gases rich in carbon monoxide (CO) and hydrogen (H_2) also contain carbon dioxide (CO_2), unreacted steam and methane (CH_4). Ash is extracted from the bottom of the gasifier in a dry or agglomerated or slag condition.

Three types of gasification processes widely used for commercial application are moving bed, fluidised bed and entrained bed gasifiers. High specific energy output is achieved in entrained bed gasifier by using fine coal particles of less than 0.1 mm, gasification temperature of between 1200 deg. C and 2000 deg. C and slagging of the ash. For a given diameter, entrained bed gasifier produces three times more product gas compared to dry ash moving bed gasifier. This has a major influence on cost of the gasifier. Fluidised bed gasifier operates in the temperature range of 850 deg. C and 1000 deg. C with the coal particles size below six mm. Due to its medium gasification temperature either air or oxygen can be used.

Cleaning

A majority of the proven gas-cleaning processes employ aqueous solution or low-boiling organic reagents at a relatively low temperature. However, until the high temperature and pressure, dry, gas-cleaning system is commercially available and introduced in IGCC, one has to bear with 1.5 three per cent loss in overall thermal efficiency. Low temperature gas cleaning necessitates heat recovery of raw fuel gas, prior to sending it for gas cleaning. This increases plant complexity. In large IGCC plants, overall thermal efficiency with wet gas cleaning is expected to be over 40 per cent.

Indian coal used for power generation contains ash around 40 per cent and sulphur up to one per cent. This difference in coal quality shifts the emphasis for development of IGCC for the use of low-grade, high-ash coal. IGCC with air blown fluidized bed gasifier offers a techno-economic viable option for the low-grade as well as high-sulphur coals.

Small Hydel Projects

It is proposed to set up 1000 small hydel projects (SHP) by the end of the eighth five year plan, of which 200 will be in the northeastern region. Efforts are made to involve the private sector in the programme, though the project can be set up through State Electricity Board or joint ventures as well.

A potential of nearly 10,000 MW has been identified for SHP PROGRAMMES but creation of 600 MW of capacity was proposed for the Eighth Plan.

It is also proposed to launch a programme for free distribution of light weight or portable micro-hydel units in the selected areas. This envisages active involvement of the local communities, non-government organisations and voluntary agencies in the setting up of such sets and subsequent operation and maintenance activity. It is proposed to distribute at least 100 units during the remainder of the Eighth Plan, of which 30 will be distributed during the current financial year.

Animal Power

Regarding draught animal power, 72 million bullocks and 8 million buffaloes can together make available 40 million horsepower providing 50,000 million units of energy per year. It is estimated that energy produced would be of the value of Rs. 10,000 crore and lead to savings of six million tonnes of petroleum annually valued at Rs. 4000 crore.

The outlay for these programmes is sought to be raised from the existing level of Rs. one crore to Rs. 100 crore which is expected to help in rural development, increasing employment, poverty alleviation and more efficient use of animal assets.

Harnessing the Gas Turbine

A newly developed high speed alternator will help a portable 20 kg turbo-generator, running on anything from petrol to coconut oil, produce an output equivalent to that of a diesel generator weighing a tonne — about 50 KW.

Engineers at Imperial College, London, state the new alternator represents breakthrough in harnessing the gas turbine's high rotational speed typically 100,000 rpm, a speed impossible to step down using a conventional gear-box.

The alternator, based on rotating discs and a series of fixed coils, paves the way for a portable generator set which is environmentally friendly, easy to maintain and cheap to mass produce.

This way achieved by turning conventional generator design. Instead of rotating an electrical coil in a stationary magnetic field, they designed a high speed alternator which has rotating magnetic discs in a series of fixed coils. This eliminates the need for rotating electrical contacts, but also creates some tough mechanical problems.

The rare-earth magnets carried on the rotating discs are low in tensile strength but are capable of generating intense magnetic fields. The magnets need to be held in position and protected against tensile stresses which are developed when the alternators are in operation. An advanced carbon fibre composite retainment shell with a specific strength some fourteen times that of high strength steel is used to make the design viable.

The new turbo-generator may be scaled to give power outputs in the range of 20 to 250 KW. Attractions over the conventional generator include its phenomenal power to weight ratio, low capital cost, multiple fuel capability including waste gases and bio-gas, low maintenance and low emissions and noise.

Producing Electricity from Fuel Gas

Based on the principle that almost any fuel can be burned more cleanly and efficiently if first turned into a gas, engineers are developing ways to produce electricity with fuel gas made from coal, wood chips, sugarcane leftovers — even sewage sludge. The process employs systems known as gasifiers, which can cook organic material into a fuel for gas turbines, the most efficient electricity generators available. While utilities currently burn natural gas in their turbines, coal is more plentiful and may eventually be cheaper.

Environmentally-Friendly Technologies

Extracting more electricity from a tonne of coal, by burning it as a gas in a turbine instead of as a solid in a boiler, would make it easier to remove conventional pollutants, like sulphur dioxide, which causes acid rain. Possibly more important, it would also mean less carbon dioxide, 'greenhouse' gas associated with global warming.

But energy experts predict that another approach may become even more environmentally important: gasifying a "renewable" fuel like wood. In the developing world, particularly, using wood or crop wastes in utility power plants could be a key to electrifying less-developed nations and making better use of forests than burning trees as firewood.

True, burning a tree for fuel releases carbon dioxide into the atmosphere. But it also opens up space for the growth of a new tree, which will absorb carbon dioxide as it grows. By contrast, digging up and burning coal — the repository of carbon accumulation — neither encourages the formation of new coal nor recycles the carbon dioxide released into the atmosphere.

Turbines running on wood or other organic matter — known generally as biomass — "could be part of a systematic response to global warming.

Fuel Requirements

Most undeveloped countries burn wood at rates far faster than forests can be regrown. A 30-megawatt plant, could use 27,000 acres of tree farms to supply it into perpetuity. That assumes the harvesting of an average of 5.5 tonnes of wood from each acre each year, and the planting of new trees.

Principle of a Gasifier

A gasifier works by heating solid fuel with steam in the presence of a carefully regulated amount of oxygen. "Dirty" solids like coal, which is filled with impurities, or solids with a low energy density, like wood or crop wastes, can be gasified, producing hydrogen, carbon monoxide, methane and other combustible gases.

Utility companies in industrial nations are increasingly turning to gas turbines, sometimes paired with steam turbines that use the gas turbines' waste heat to make even more electricity. These are replaying old-fashioned generating stations that burn fuel in a boiler furnace to make steam and then run the steam through a turbine to make mechanical power, which turns a generator.

Conversion Efficiency

Burning a fossil fuel like coal in a boiler a generating station converts, at most, only 32 per cent of the fuel's heat into electricity, the rest of the heat energy escapes up the stack or into a cooling tower or condensor. The best gasturbine plants, by contrast, are 53 per cent efficient.

A high temperature system cleans up the gas while it is hot. The gas is spun in an artificial cyclone, whose centrifugal force throws off particles, including sodium and potassium, two metals that can ruin the blades of a gas turbine. In a separate vessel, an absorbent material — zinc ferrite or zinc and titanium dioxide — absorbs the sulphur.

In addition to improving the energy properties of coal, gasification might turn useless agricultural waste, like the stalks of harvested cotton plants, into valuable fuel.

Power from Biomass

A national programme is being launched for bio-energy utilisation relating to co-generation and recycling of urban and municipal wastes to save imported fossil fuel costs.

About 600 MW of power was to be generated through the use of biomass during the Eighth Plan.

Technology is available to tap 3000 to 4000 MW of surplus power generation potential in 400 sugar mills through the use of optimum co-generation schemes. Improving energy supply to rural areas through installation of solar photovoltaic, biomass gasifiers, small hydel and wind energy systems for power generation are being attempted.

Solar Hygen System

A Rs. 124 crore project, using hydrogen for generating 500 KW of electricity from solar energy, by the Rajasthan Electronics and Investment Limited (REIL). Christened the 'solar-hygen system' is under consideration.

Hydrogen, as a source of energy, can meet the demands of both the consumer and the industry sector. The two factors to be kept in mind are the perennial water supply and an open space away from human habitation, since the photovoltaic cells can be tampered by the locals.

An agency should be created exclusively to cater to the needs of the solar-thermal, photovoltaic, biomass and wind mills, and be responsible for generation of power also.

Designs for Offshore Power Stations

As per estimates, about 40,000 MW of power can be generated from wind, about 50,000 MW from waves, currents and tides and about 50,000 MW from ocean thermal energy conversion (OTEC), all of which are favourable to environment. While most efforts are concentrated on designing land-based systems, harnessing energy from wind, wave and surface currents for generating electricity out at sea without affecting marine environment is feasible.

India has a coastline of about 6,000 km making available plenty of wave and surface current energy and is also wellplaced as far as solar energy. It is therefore imperative that these inexhaustible non-polluting sources of energy be fully exploited for generating electricity. About 40 MW of solar energy and 38 MW of wind energy have been converted into electricity. There is a vast potential of the order of 80,000 MW or more of wind, wave and surface current energies that are yet to be tapped. India has made considerable progress in the installation of wind power generators and wind energy farms. A capacity of 38.31 MW has been completed and 24.47 MW capacity is under installation.

Wave Energy

For generating electricity from wave energy, three basic methods have been reported. One of these uses a Well's air turbine-generator combination mounted on top of concrete caisson which acts as the breakwater. In this scheme the energy in the waves is imparted to column of air inside a caisson. This air drives the Well's turbine which in turn drives an induction generator. Based on this principle, an experimental 150 KW wave energy plant has been installed.

The second method for generating clean electrical power is based on air turbines actuated by air movement continuously generated by the oscillating movement of a flexible membrane. This membrane oscillating under wave action produces a constant air movement in the space between the flexible membrane and solid hull which drives the well's turbine. The turbine is used to drive the generator.

The third method is plunger type reciprocating mechanism which moves inside a moon pool when activated by the waves. The reciprocating motion is converted into a rotary one and through a system of gears drives the generator. Excepting for the flexible membrane used in the second method, the first and the second methods of generating electricity form wave energy are similar. The second can be chosen where a total isolation of water from the column of air that drives the turbine is desirable. The third method, though equally simple as the first two is yet to be tried out in real-world. The final choice of a method will, however, depend upon the simplicity, cost and ease of maintenance offshore.

Ocean Energy

Ocean currents are of many types. Drift or wind currents are those developed from friction exerted on the sea surface by steadily blowing wind. Slope currents are caused by the inclination of the watersurface developed by wind, barometric (atmospheric pressure) differentials or inflow from rivers. Density currents are caused by differences in the density of water. Tidal currents are caused by tidal actions. Currents are also caused by internal waves propagated along surfaces of large density discontinuity within the body of the sea.

The average ocean currents are composed of large scale circulation of water due to the distribution of mass in the sea which is developed by heating at the tropics and cooling at the poles. This heating and cooling process induces a general surface motion towards the poles and a depth motion towards the equator. The motions are further modified due to the effect of the Earth's rotation (coriolis force), gravity, irregular solid boundaries of the sea, and by wind effect. In contrast to other small scale, more or less local currents, the large scale circulating currents transport great amounts of water and the flow is laminar.

Energy for power generation can be tapped from the ocean currents either by using a device which works on the principle of a horizontal axis current meter or by using a suitable slow speed turbine.

Non-Conventional Energy Sources

India faces power shortages year after year. This can be made up by installing more fossil fuel-based power plants. This would, however, give rise among other things, to two problems — further depletion of the coal and oil reserve and degradation of the environment due to pollution. Fortunately, India has tremendous potential for harnessing the abundantly available non-conventional energy resources in the form of solar, wind, wave and ocean current energies. The right technology and talents are there to tap these inexhaustible resources for meeting India's power requirements.

Nuclear Energy

The existing capacity of the four operating nuclear power plants is 1465 MW. With the commissioning of the four plants now under construction, the generating capacity will be increased by another 2585 MW as follows : the second unit of Narora (235 MW), Kakrapar in Gujarat (2×235 MW), Kaiga in Karnataka (6×235 MW), Kota (third and fourth unit of Rajasthan atomic power plant) — (2×235MW).

The third and fourth unit at Tarapur would have capacity of 500 MW each, while another nuclear power plant (2×500 MW) was also proposed to be established to be 2000 AD to achieve a target of 6050 MW of nuclear power in the country.

India is lagging behind many countries in the generation of nuclear power. The nuclear share of electricity generation in India is 22 per cent. This compared poorly with France (75 per cent), Belgium (60 per cent), Hungary (51 per cent), Republic of Korea (49 per cent), US (20 per cent and USSR (12 per cent).

II. Energy Management

The Nation can save 30,000 MW of power and Rs. 1,35,000 crores in investment by the end of the 10th Plan through effective management of power demand and energy conservation.

By 2000 AD the peak power is expected to rise to 1,44,300 MW and energy demand to 824 billion units.

This would require enhancement of the installed capacity to 2,11,300 MW from the present 72,300 MW. The funds required to bring about additions of such magnitude is a staggering 6,25,000 crores. The costs could be reduced through demand side management.

Resource constraints had already limited the magnitude of capacity additions and the country would face a severe shortage in demand and energy by the end of the eighth Plan and beyond.

Resource constraints had already curtailed the capacity addition in the eighth Plan and the financial outlay for the power sector to about Rs. 79,589 crores. Estimates indicated that with this level of outlay, the capacity addition could be only about 25,800 MW, including 2,800 with the help of private sector participation.

Therefore, there is greater need for power utilities to improve their operational efficiencies such as better productivity of installed capacities in generation, transmission and distribution, improved arrangements for the transfer of power between regions, besides the renovation and modernisation of existing power plants.

India's per capita power consumption of 300 units would rise to only 600 units by 2007 AD. While there is need to enhance the consumption, the attempt should not be to reach the profligate level of 15,000 units per capita in advanced countries but to raise it to a level which would enhance the quality of life of the common man.

The concept of demand side management sprang from the concerns about the environment and focus on energy conservation. Traditionally, electric utilises responded to increased power consumption by building new generating plants.

Utility executives were sceptical about the effectiveness of demand side management when it was introduced as they felt it would divert resources away from much-needed new plants. Demand-side programmes in the USA helped in cutting the potential peak demand in 1991 by 26,700 MW and annual electricity use by 23,000 Gwh (Gega Watt hours).

Power Delivery System of the Future — Flexible AC Transmission

The demands of the twenty-first century, with increased growth and interconnection of power grids, will stretch existing delivery systems to their limits and create new requirements for flow control, system stability, and capacity enhancement. Flexible ac transmission system (FACTS), based on advanced solid state controllers, incorporate the capabilities that will be needed to reduce the susceptibility to power disturbances which occur often. FACTS controllers offer tremendous speed, precision and reliability improvements over the mechanical switching devices they will replace, and they can be added onto an existing system individually as they become available. Full application of FACTS technology in the coming decades promises to truly transform the power delivery system.

Since the earliest days of electric power, engineers have wrestled with a basic conundrum: although high-voltage alternating current facilities low-loss transmission over long distances, it is difficult to control. The new devices, which use large solid-state switches called thyristors, will help utilises transmission system capacity while reducing susceptibility to power disturbances and enhancing the control of flow. Three kinds of thyristor-based controllers are now being incorporated into utility demonstration facilities in the U.S.

Advantages

In high-power applications, thyristors enjoy two great advantages over the mechanical devices — circuit breakers, transformer tap changers, shunt capacitor switches, and the like that now control the ac power system. First they are much faster than the alternatives and can redirect power in a fraction of a cycle. As a result, these devices will be able to perform functions, such as counteracting transient disturbances on a transmission line, that could not even be contemplated with mechanical controllers. Second, mechanically switches devices tend to wear out and thus are generally operated sparingly. For example, a transformer used to compensate for shifting load is generally limited to less than a dozen motor-driven tap changes a day — with a full range variation requiring more than a minute to complete. Thyristor-based controllers, on the other hand, can switch power twice each cycle without wear.

Thyristor controllers can provide unprecedented opportunities to regulate ac transmission, increasing or decreasing power flow on specific lines as desired and responding almost instantaneously to nascent stability problems. The result, known as FACTS, will enable utility networks to operate like giant integrated circuits — to be more highly automated circuits — to be more highly automated, self-correcting, reliable, and productive.

Economic Benefits

In addition, greater flexibility will mean that utilises can operate their transmission systems closer to inherent thermal limits (the point at which lines literally become hot enough to sag too much) rather than adhering to the lower rating on many lines imposed by concern about system stability or about overloading one of the parallel paths. A typical 500-KV line, for example, might have a rating of 1000 MW, compared with a 3000 MW thermal unit.

The potential economic benefits of increased loading of existing transmission lines are both varied and substantial. In the first place, building new lines can be much more expensive than upgrading existing lines — if new ones can be built at all in the face of mounting environmental opposition and the difficulty in obtaining new right-of-way.

In many cases, though it may be necessary to combine the application of FACTS technology with the upgrading of existing lines or the construction of new lines.

Large savings can also be realised from reduced generation reserve margins — the generating capacity needed to serve as backup rather than to meet actual demand for electricity. By using FACTS controllers to strengthen transmission ties between various control areas and power pools, utilises can share generation reserves more productively and may thus be able to reduce reserve margins.

Increased capacity FACTS controllers allow transmission lines to be loaded closer to their thermal limits, in some cases actually doubling their capacity. This can significantly reduce the need for building new lines.

A thyristor-controlled series reactor can be used on transmission lines that requires rapid load reduction or limiting of fault current. The use of these devices together may in some cases provide a lower-cost alternative to a phase-angle regulator. Design of the series reactor is scheduled to begin in the mid-90s after study of potential control strategies and ways of minimising hardware costs.

A thyristor-controlled voltage limiter can protect utility equipment from over voltages and thus permit system operation that would not otherwise be possible. Over voltages may result, for example, from capacitor switching or from the dropping of a large load from a heavily compensated lines.

Automation of Distribution to Reduce Power Loss

Automation of power distribution system, with the advances made in technology, would help optimum utilisation of the limited resources.

Automation would ensure efficiency in feeder control especially in low tension supply and minimise equipment failures. Problems like theft of energy could be solved when meter readings were done at the sub-station also, and the billing was computerised.

Very Low Consumption

The cost of power was comparatively higher in the country and the per capita consumption which was only 250 units was much less compared to 2,500 units in some of the least developed countries. The cost of power generation capacity which was only Rs. 50 lakh per MW has increased to Rs. 2.5 crores.

Distribution automation was accorded least priority so far despite the fact that much of the power loss occurred in distribution. Due to lack of demand for the automation technology, industry also could not catch up with the advances made in the sector. Practical experience only would help the distribution system orient to the new technology.

Automation was a necessity not only because of its cost effective benefits but also from the environmental angle. Recent advances in computer and communication technologies would help automation of distribution.

Encouraging Signs on the Power Front

The plant load factor (PLF) of the thermal power plants is slated to rise by one per cent. The increase in PLF will yield additional generating capacity, leading to a growth in revenues.

PLF will cross 58 per cent, a marked improvement over the last three years as it was only 53 per cent in 1990-91. Even the one per cent increase will mean an additional availability of 723 MW of power generation of which investment costs are currently estimated at about Rs. 2,500 crores.

The huge losses being incurred on subsidised power supply is evident from the fact the generating cost is estimated at Rs. 1.22 per unit of electricity while the actual recovery is only 13.5 paise from the agricultural sector. Agriculture, at the same time, is among the largest consuming segments with as much as 27 per cent of the country's available power, being used by this sector.

Efforts are being made to reduce the huge transmission and distribution losses, estimated at present to be 22.8 per cent of the total power generation. Along with the expected increase in PLF the proposed reduction in transmission and distribution losses will help the country gain additional power generation of about 1450 MW.

Measures to increase the availability of power to various consumers, includes increases of plant load factor (PLF) by at least one per cent per year in the case of those with 60 per cent, PLF, by two per cent per annum in the case of those with 40 per cent PLF and by three per cent every year in the case of the low performing SEBs, particularly those with less than 40 per cent PLF.

A one per cent increase in the PLF made available 500 MW of power and with the national average targeted to go up from 56.6 per cent PLF at present to 60 per cent by the end of the Eight Plan period, at least 1,500 MW of additional power would be available. Cutting down on transmission and distribution losses from 22 per cent to around 17 per cent by the end of the Plan period, would result in additional revenue generation of around Rs. 11,500 crores since a one per cent reduction in such losses could yield Rs. 2,300 crores.

National Power Grid

The national grid could be a reality once the high voltage transmission was upgraded from the present 400 KV level to 1000 KV with minimum transmission losses. The concept of national grid was more viable than the existing system wherein problems had cropped up due to break down in power supply.

Power Grid Corporation

Power would be systematically supplied to various States with the establishment of the Power Grid Corporation of India. The Power Corporation, estimated to cost Rs. 7,000 crores, was to be set up in the present plán. Though the project would take long time, it would solve the power crises gripping the nation.

People should be educated on how to conserve energy. The project to generate power from sea water is being studied. If this proved successful, the whole country would adopt this method of hydro and thermal power generation.

The transmission loss in the country is about nine per cent during non-peak hours and 20 per cent during peak hours.

There is need for reducing wastage in transmission, Rs. 12,000 crores was needed for transmission improvement.

There would be an additional generation of 30,500 MW at the end of the Eight Plan, requiring Rs. 1 lakh crores, out of which the Centre and the States would fund Rs. 80,000 crores. The balance would come from private sources. Despite this, the country would still be deficit in power, as the demand is going up by leaps and bounds. Originally, urban and industrial power needs were more than the rural requirement. Today, it is reversed with the rural and domestic demand touching 52 per cent and the urban and industrial demand 48 per cent.

Plant Load Factor

The plant load factor in almost all the power plants was very low at 50-52 per cent, while in the West it was 75 per cent.

If we can increase the plant load factor and minimise the transmission losses, power shortage can be considerably reduced.

Competitive Bidding for Power Projects

Investment proposals, worth over Rs. 1,00,000 crores for power projects in the private sector were received and "competitive bidding" for such projects and also "global bidding" for equipment supplies, are being considered. Competitive bidding is to be introduced on a prospective basis and will not include projects already cleared in the private sector.

Only these projects which had not been cleared would come within the gambit of the scheme.

There are 75 proposals for setting up power projects in the private sector. Of these 35 would help create additional capacities of 32,000 MW involving an investment of Rs. 1 lakh crores. The proposals were from consortia of leading multinationals and Indian corporations.

III. Energy Conservation

India's per capita energy consumption though very low compared to global standards, has been rising rapidly. It doubled from 110 to 219 kgoe (kilograms of oil equivalent) between 1970 and 1994. The energy demand has consistently outstripped the supply, causing permanent shortage of petroleum products, electricity and quality coal.

The country has witnessed a sizable increase in the production of commercial forms of energy. Between 1960-61 and 1990-94 the coal production increased from 55.67 million tonnes to 211.73 million tonnes, the lignite output from 0.05 million tonnes to 14.07 million tonnes, crude oil from less than a million tonne to 33 million tonnes, hydro electricity from 7.84 billion Kwh to 71.54 billion Kwh, thermal power from 8.1 billion Kwh to 186.5 billion Kwh and nuclear generation to 6.24 billion Kwh.

Biomass forms of energy still accounted for nearly 40 per cent of the total energy consumed in the country. Nearly 90 per cent of India's rural population still relied on fuel wood, animal dung and vegetable waste to meet its cooking needs.

Despite the increasing investment in the energy section, India's ability to meet the rapidly growing demand for energy has not kept pace with production. In the First Five Year Plan, less than 20 per cent of the total outlay was allocated to commercial energy sector, which rose to 21.2 per cent respectively in the Fourth and Fifth Plans, to 28 per cent during the Sixth and Seventh Plans and slightly down to 26.28 per cent in the Eighth Plan (1992-97).

On the assumption that energy intensity of the GDP would register a yearly compound growth of five per cent, the total commercial energy demand has been estimated at a staggering 600 mtoe (million tonnes of oil equivalent) by the year 2010 as against the present level of around 200 mtoe . In less than two decades, the demand for commercial energy will expand by three times. For instance, the coal requirement has been projected at 620 million tonnes, oil at 175 million tonnes and natural gas at 175 million cubic metres a day. If the projected demand is to be met the country will have to import 200 million tonnes of coal and 116 million tonnes of oil. The primary electricity demand is also expected to rise to 217 Twh by the year 2010.

If India were to plan for energy supply and use it in a rational manner it will have to incorporate the following options (i) demand management (ii) efficiency improvement (iii) optimal supply/demand mix (iv) accelerated utilisation of renewable energy and (v) series of incentives and disincentives for energy conservation. This way, the energy demand in the next two decades could be brought down. Economic potential for cogeneration of renewable energy in major industry sectors in India is 1500 MW. A study showed that the industries with the highest potential were fertilizers, basic chemicals, refinery and pharmaceuticals. At present, natural gas is being made available to only fertilizer and petrochemical industries. If gas were to be made available to all industries, the potential for cogeneration could be 3000 MW in less than a ten-year period in two States alone.

But the natural gas that comes out with crude is at present being flared for lack of infrastructure to capture and use it. Recently the Government launched a Rs. 7,200 crore World Bank aided scheme to reduce the flaring to zero by 1994-95. The Petroleum and Natural Gas Ministry has planned to implement seven mega oil production projects that may also double gas generation at 95 million cubic metres. If more gas based units including power stations could be concurrently set up, then pressure on oil thermal power, coal etc., will come down.

While consumption could be restrained immediately, statistics reveal that after the initial depression the demand curve once again continues its upward trend.

The process of economic liberalisation initiated cannot be sustained unless a national energy policy is framed. Besides energy conservation it should also bring out the need for immediate development of non-conventional sources with renewable energy options. At a petro products consumption level of 22.16 million tonnes, industry accounted for 48 per cent, transport 33 per cent, domestic 15 per cent, agriculture one per cent and others three per cent. By 1990-94 when the consumption rose to 55.46 million tonnes, the transport sector topped with 45 per cent, industry 35 per cent, domestic 16 per cent, agriculture one per cent, others 3 per cent respectively. The need exists for energy conservation with a set of incentives and disincentives to cover the obsolete but still functioning industrial units.

The main source of non-conventional energy in the country continues to be forest or fuel food accounting for nearly 65 per cent of its type.

India happens to be the largest importer of kerosene. According to an estimate nearly 75 million Indians are exclusively dependent on kerosene for cooking and lighting. If bio-gas could be provided on a massive scale, there could be a sharp reduction in kerosene consumption.

The ratio of thermal-hydel mix in power generation rose from 60:40 at the end of the Fifth plan to 72:28 in 1990-94. While thermal is an exhaustible source, hydel is not. But because it takes a longer period to be set up, the thermal is always preferred. Latest technological development has shown that it impossible to set up hydel power stations with a lower gestation period. A recent survey has shown that against a national potential of 600 billion Kwh of hydro power, only 18 per cent of it had actually been exploited. Yet another official survey showed that mini and micro hydro stations alone could yield 5000 MW of power.

Thus, exploitation of non-conventional energy like wind, solar and biomass could yield 17,000 MW through use of bio-gas plants by 40 million persons, 5000 MW from wind energy. Strangely, in the Eighth plan, we intend to generate only 100 MW from wind energy.

IV. Conservation of Energy in a Process Industry

In a typical paper Mill, the above is achieved by (1) Estimation of the running load; (2) Reduction in loads; (3) Maximum demand control; (4) Steam wastage control; (5) Monitoring Power factor; (6) Checking leakage in steam pipes; (7) Reducing the size of water hoses; (8) Forming an Energy cell to supervise.

Other Industries

(1) Energy efficient motors selection; (2) Varying their application as per the load; (3) Reducing hours of operation and kilowatt energy reduction; (4) High efficiency Motors for improved power factor, lower temperature rise, lower noise, higher insulation life and capacity to operate wide voltage fluctuations, reduced pay back period, greater cost-benefit ratio.

POLLUTION-FREE PULPING TECHNIQUES

Following are the technologies which may revolutionise the pulp and paper industry.

1. A radically new technology involving alcohol and water 'organo solve' pulping substitutes ethanol for traditional sulphur containing pulping chemicals. Mills based on this technology will be free of the odour normally associated with kraft pulp mills and will have low volumes of effluent per metric tonne. And with no sulphites and sodium hydroxide, by-products such as lignin and pentoxes sugars are purer than those from the kraft process.

The process offers valuable by-products such as lignin furfural, ethyl acetate and acetic acid.

2. *Using urea* : In developed countries, ammonia pulping has been tried in production of high yield chemical pulp for hard woods. Though results were encouraging, it could not be commercialised. Alternatively, urea, again nitrogen based which gives ammonia on decomposition, was also tried with an added advantage of disposing the pulp effluent to agricultural fields as a source of fertilizer. Urea can be tried as a pulping chemical for agricultural residues and possibly later for small paper mills.

As most small mills are closer to fields, the pulping effluent generated can be easily disposed as a source of fertilizers without adding any effluent load.

The chemi-mechanical pulp of rice straw produced by eight per cent urea was comparable with that by four per cent sodium hydroxide. As urea costs one-fourth of sodium hydroxide, the total cost of chemical per tonne of pulp produced is half for urea pulping. With urea pulping there is little pollution load as the effluent generated can be sent to the fields to benefit from the nitrogen content.

3. While these two technologies are being commercialised, bleaching pulp using oxygen is being improved.

The darkest effluent contributing to BOD & COD is generated in conventional bleaching processes using cholorine-caustic-Hypo-Hypo sequence (C-C-H-H) formation of dioxins as a result of bleaching has been worrying environmentalists. In this context, oxygen bleaching assumes important one.

Most integrated paper mills making bleachable grade cultural paper adopt conventional CCHH bleaching for a brightness level of 78 ISO. This paper has a tendency towards colour reversion due to over degradation in hypostages, which also affects paper strength. If 85 ISO is desired the most viable sequence must be selected which can improve brightness without affecting strength and leading to lesser pollution load. The best way is to incorporate oxidative extraction (E/O) stage in the bleaching sequence. This can be adopted in a running plant by retrofitting of "E" stage with "E/O" stage with few modifications.

A dissolving grade pulp mill has developed and implemented an oxidative extraction scheme with in-house R & D. The E/O stage has been retrofitted in a 200 TPD pulp mill encompassing the entire system of oxygen bleaching inclusive of hp-pump, reactor, washer etc. The strength and brightness of pulp can be improved by the "chlorine dioxide", (D) stage in bleaching sequence. The technology has been perfected and is being operated.

Advantages are :

(a) A generation of a unobjectionable effluent.

(b) Reduction of BOD 5, COD by 50 per cent and colour by 65 per cent.

(c) Toxicity load in effluent is slashed.

(d) There are large savings in recurring treatment.

CLEAN PULP MILL

4. A technological system that completely eliminates the use of chlorine in the bleached kraft pulping process has been developed. Dioxins and furans have been linked to the chlorine used in the standard bleaching process. The technologies used in the mill will virtually eliminate the production of dioxins and furans.

Upgraded pulping process reduces biological oxygen demanded (BOD) from three kilograms per ADMT to one kilogram per ADMT, thereby decreasing the total amount of by-products discharged by the mill into the river.

The level of absorbable organic halides (AOH) has been cut from 1.3 kilograms per ADMT to 0.35 kilograms per ADMT, thereby reducing the total amount of chlorinated organics discharged into the river.

The amount of discharge from this process will be lower than from any other bleached kraft mill.

Selection of the site is based on a number of factors, including quality of the soil, proximity to major towns, environmental impact, restrictions on effluents and infrastructure, water, power and natural gas requirements.

Forests supplying the raw material are managed according to three principles (1) integrated resource management (2) sustained yield and (3) maintenance of a high-quality environment. Wood will be harvested from the surrounding region where sufficient volumes are available to sustain the pulp mill and any future expansion on a long-term basis.

9. Policies and Strategies (National and International)

ENVIRONMENT AND DEVELOPMENT

Industrial pollution monitoring, control and safety should receive the greatest attention.

The conceptual confrontation between development needs and environmental consciousness should end and if a balance was struck between these two concepts, a confrontation could be avoided, ''If we apply environmental requirements too rigidly and mindlessly, confrontation will occur.''

In order to strike an ecological balance, ''We will have to compromise on concepts such as economies of scale''. In Czechoslovakia for e.g., power needs had been met by small and medium hydel plants generating between 50 and 100 MW.

CONCERN OVER POLLUTION

Permission should stop and giving for setting up more industries in pollution and in order to check further pollution.

The existing industries in these areas were discharging toxic effluents into river, lake and other reservoirs and tanks.

The pollution caused by these units was so heavy that groundwater in several adjoining villages was becoming contaminated affecting drinking water supply and agriculture. Erring industries should pay compensation to the affected farmers. While development was essential, it could not be at the cost of human life, did not appear to be conscious of the effects of pollution caused by units.

The closure of certain inudstries but allowed them to function though they had not taken any measures to check pollution. The State Government must now intervene and initiate stern measures to close down units which do not install effluent treatment plants.

ENVIRONMENTAL CLEARANCE APPROVAL A MUST FOR NEW UNITS

Environmental clearance is to be made compulsory for all new industrial units and for any expansion or modernisation of existing units.

Development activities have had an adverse impact on land, water and air and natural resources are being destroyed and depleted. This is threatening the very survival of not only plant and animal species but also human beings.

It is compulsory for any project to be located within 10 Km. of a reserved forest or a designated ecologically sensitive area to get clearance. This will also apply to projects located within 25 Km. of the boundary of all National parks.

Site clearance will be necessary for mining projects and for thermal power plants and hydel projects as these lead to largescale felling of trees, drilling and digging.

Among the industries concerned are atomic and thermal power, fertilizers and pesticides, petrochemicals and plastics, drugs and pharmaceuticals (except formulations), various chemicals, integrated steel plants, multi-purpose river valley projects, synthetic fibres, and rubber, also ceramics, glass, engineering, pipelines, soaps and detergents, paper products and milk processing and paints. For cement, steel, sugar and textiles, production capcities are to be specified.

POLLUTION REGULATIONS COMPLIANCE

Manufacturing process should not be harmful to environment in any way at any of the stages. Companies should comply with the pollution control rules and they should provide at least proof of intent to adopt pollution control technology within the stipulated time to obviate legal action.

Fiscal incentives should be provided for waste control technology adopted by the companies to encourage a clean environment. The stress was on clean technology to control pollution and not on pollution management as the latter did not arise if the former were to be taken care of.

The air quality had reached a dangerously critical level in respect of suspended particulate matter, like sulphur dioxide and nitrous oxide levels.

The groundwater pollution too is at its worst high with fluorides, nitrates and zinc found in large quantities. Industries contributing to the alarming situation have been identified and measures should be initiated to curb the pollution levels.

The automobile manufacturing units should produce better engines to curb air pollution and petrol and diesel wastage.

ENVIRONMENTAL AUDIT

An audit of the pollution control methods in the factories would not only help improve the efficiency of the companies but also dramatically curb the accidents.

As regards the air and water pollutions, the chemical industry generated about 500,000 tonnes of wastage consisting toxic material every year. Though benzedine had been identified and banned there were many more to be banned. The usage of appropriate technology for pollution control, would help the industries to minimise waste load too.

There have been several instances of abortions and formation of deformed foetuses among women due to pollution. As for the vegetation, large acres of fertile land was rendered useless, while major lakes in the area turned unfit with the effluents being dumped into them.

The measures suggested include installation of air monitoring equipment in each chemical industry, display of data of air and water pollution levels gathered from chemical industries.

COSTS OF POLLUTION CONTROL

Environment concerns were being used by the developed countries to insist on reallocation of budgetary priorities, cuts in military spending, raising of price of energy, and encouraging resource flows to the private sector. In the name of future generations, the developed countries were not taking fully into account the present imbalances in global patterns of production and consumption. The promotion of economic growth in developing countries was essential to address the problem of environmental degradation.

As far as ozone depletion is concerned, the pattern of global emissions of CFCs tell almost the whole story. Of the total estimate of 1.2 million tonnes of emissions of CFCs 300,000 tonnes are emitted by the United States alone, India's share is only 6000 tonnes a year. All the developing countries together add up to five per cent of the total emissions. And even these have occurred in recent years while the industrialised countries have been emitting CFCs for the last five decades.

The scientific data on global warming and emissions of greenhouse gases points guilty finger at the developed countries, which is at the top of emitting carbon dioxide, a major greenhouse gas. An estimate made by the World Resources Institute put the global per capita emission of carbon dioxide from burning of fossil fuels, gas flaring and cement manufacture at 1.7 tonnes for 1987. That year per capita emission by the U.S. stood at five tonnes while the figure for India was 0.4 tonnes. The average for Europe was calculated as 2.4 tonnes per capita.

The total global carbon dioxide emissions were estimated at 1.6 billion tonnes in 1950, 5.26 billion tonnes in 1980 and about 7 billion tonnes now. Some scientists are predicting that even at slower increase rates by the year 2000 the world carbon dioxide emission could be over 20 billion tonnes. This gives us an idea of the global warming that would take place if this trend remains unchecked.

The responsibility of countries must be in relation to the damage caused to the environment. All incremental costs incurred by developing countries in phasing out technologies using ozone destroying chlorofluorocarbons should be borne by the developed countries.

In the non-industrialised world which has not appreciably engaged in the buildup of greenhouse gases, three distinct groups have merged—the oil exporting countries, the small island states, and the third group which includes large populous states like India and China.

Recently, the Government of India made its stand very clear when it stated that the long term global objective must be to bring down emissions in the affluent nations where they are high and allow the developing countries to increase their per capita emissions till the two converge at a common point to be fixed with the help of scientific data.

Also the polluter must pay in principle. Since the developing countries are in no way responsible for the state of the environment today, any costs for adopting new technologies to cut back on carbon dioxide emissions must be paid for by those who have caused the problems.

Developing countries have also taken the stand that they would not sign any forest or bio-diversity convention that infringes their sovereignty. The green movement has become increasingly important and vocal. The recent discovery of a hole in the ozone layer in the atmosphere, has already created a fear psychosis in the affluent countries.

A recent issue of 'Climate Convention' brought out by Greenpeace has pointed out a new and dangerous concept of carbon colonialism.

What could be acceptable to the developing countries is the fixing of a carbon emission quota on the basis of current populations and on a per capita basis, with those who emit more paying to those who emit less for using their environment space.

The question of growing populations of countries like India and China can be overcome if the norm of environment space is accepted on the basis of the current populations and fixed at that point.

The ecological imperative underlines the urgency of an altogether different and radical solution, the adoption of a model that can co-exist with the inexorable laws of nature.

THE CRUX OF THE PROBLEM — THE ANSWER LIES ELSEWHERE FOR A JUST SOLUTION

Although there is an alround effort to clean up air and water sources in the cities to reduce emissions from automobiles and to make them more energy efficient, to limit energy consumption in general through the use of more efficient technologies, and to evolve newer technologies and control mechanisms for reducing and checking air and water borne effluents from indsutry, even by shifting some of the more persistently polluting industrial activities elsewhere, it is no where to solve the problem.

The effort is to clean up the immediate micro-environment of the cities and smaller habitations.

The level of air pollution on many of the major highways in the industrialised countries with their continuous stream of high speed vehicles is likely to be lower than that on the highways of India. The bigger and much more powerful cars in those countries in fact often consume less than half the fuel guzzled by our lumbering car and consequently cause that much less pollution.

This cleaning up of the immediate micro-environment of the cities of course does not solve the problem. The cumulative load of the pollutants produced by the life-styles of the affluent is still too heavy for the earth to bear. According to current estimates, average energy consumption in the developed world amounts to the equivalent of 5,000 kg of coal per capita per year. The average for the highly industrialised countries is around 11,000 kg per capita per year. A country like India on the other hand consumes the equivalent of only 250 kg of coal per head.

At their level of energy consumption, such countries add five tons of carbon per head a year to the global atmosphere in the form of carbon dioxide, while India adds about 0.4 tons. The carrying capacity of the global environment is currently estimated at around 0.7 tons per capita per year. The large amount of carbon added annually to the atmosphere by the activities of the affluent is in addition to the undigested carbon already accumulated in the atmosphere. In the 90 years of this century the most industrialised country is estimated to have added a total load of 240 tons of carbon per living citizen, while in the same period Indian contribution to this environmental load does not amount to more than six tons per living citizen.

Carbon load in the form of carbon dioxide produced by the burning of fossil fuels is of course not the only pollution that the earth's environment has to suffer. There are also a variety of other chemical pollutants that the Industrial activities of the West add to the atmosphere. And the load of some of these chemicals, like the chlorofluorocarbons used especially in the refrigeration industry is known to have become already unbearable for the atmosphere of the earth.

This accumulated and fast accumulating load of pollutants leads to macro-environmental changes like rise in the temperatures across the globe and depletion of the protective ozone layer of the earth atmosphere. These macro changes are not currently amenable to technological solutions and these changes can prove detrimental to the continuation of affluent lifestyles. Their experience of cleaning up the micro-environments over the last decade or so, however, has given the confidence that given a somewhat longer period of, say two to three decades, they can find technological solutions to the problems of macro-environmental pollution also without having to change their life-styles in any essential way. There is confidence that with improved energy efficiencies in industry and some restructuring of the agricultural, housing and transport sectors, it may be possible to reduce the annual per capita environmental carbon load by about half without sacrificing on their mobility and comforts.

This confidence however is based on the assumption that the rest of the world shall not increase its energy consumption and consequent carbon load on the global environment to any appreciable level. It also assumes that the rest of the world shall not cause any major imbalance in the current state of the environment through, for example, too fast an exploitation of their forest resources or too abundant a use of chemicals like chlorofluorocarbons, that pose a threat to the global environment.

As a part of international environmental cooperation, they are seeking guarantees from the rest of the world that the latter will not increase their energy consumption except marginally, that they will not exploit their forest and other natural resources with the same abandon with which the former have been exploiting the resources till now, and that they will not indulge in certain type of industrial activities that are hazardous to the global environment.

On their part they must offer guarantees to reduce energy consumption to significant level or to abandon any of the activities found to be unsustainable. They must leave non-affluent countries like India to pursue their own environment-friendly ways.

No one can accept any unequal arrangements in the name of environmental protection. Countries like India should decide to avoid the affluent path of excessive consumption of natural resources and reckless pollution of the global environment, without possibly undertaking the self-denial as a sacrifice to the cause of preservation of the life-styles of the advanced countries.

If however, the industrialised countries undertake to change their ways so as to reduce per capita burden on the global environment to a level at which the whole of humanity can live without straining the bearing capacity of the earth, then of course there is a real possibility of understanding with the rest of the world. If, for example, the affluent undertake to reduce within say the next decade, the emissions of per capita carbon load from around three tons a year to about 0.7 tons that the earth can reasonably bear, then there can be a way as to how to make this possible with the minimum dislocation of life. With such a commitment of returning to environmentally sustainable life-styles, there can also be methods on allowing a somewhat larger than the average quota of energy consumption and environmental pollution, taking into account the relatively difficult natural environment of the regions of the world. But any such search can proceed only from a prior commitment to giving up the ways of life that are unbearable on earth.

The problem of unsustainable environmental pollution is a product of the lavish way of living. Now that the environmental consciousness has been aroused, the rest of the world can cooperate and help in making the necessary adjustments for returning to more environmentally friendly life-styles. The world cannot however be expected to make sacrifices in order to ensure the uninterrupted continuation of the environmentally destructive ways.

NEW ENVIRONMENTAL POLICY : INCENTIVES FOR POLLUTION CONTROL

National Policy for abetment of pollution envisages compulsory annual environmental audit for industrial concerns and a host of fiscal incentives for adopting cleaner technologies to check and control pollution.

The policy sought to shift emphasis from defining objectives for each problem area towards actual implementation and the focus is one the long-term because pollution paritcularly affected the poor.

The objective of the national policy has been defined as to integrate environmental considerations into decision making at all levels. To achieve this the steps envisaged in the policy are: To prevent pollution at source, encourage, develop and apply the best available practicable technical solutions ensure that polluter pays for the pollution and control arrangements focus protection on heavily polluted areas and river stretches, and involve the public in decision-making.

Regarding the environmental audit of the National policy, industrial concerns and local bodies should feel that they have a responsibility for abatement of pollution.

The procedure of an environmental statement will be introduced in local bodies, statutory authorities and public limited companies to evaluate the effect of their policies, operation and activities on the environment, particularly compliance with standards and the generation and recycling of waste.

An annual statement will help in identifying and focussing attention on areas of concern, practices that need to be changed and plans to deal with adverse effects. This will be extended to an environmental audit. The measures will provide better information to the public.

While regulatory measures remain essential for the effectiveness of the policy, new approaches for considering market choice would be introduced.

The aim of these "new approaches" is to give industries and consumers clear signals about the cost of using environmental and natural resources. The expectation is that market-oriented price mechanisms would influence behaviour to avoid excessive use of natural resources.

There are at present, several fiscal incentives for installation of pollution control equipment and for shifting polluting industries from congested areas. The items for which Excise and Customs rebate are allowed will be reviewed. This will stimulate the advancement of abatement technologies and create increased demands for the products.

Economic instruments will be investigated to encourage preventive measures, internalise the costs of pollution and conserve resource, internalise the costs of pollution and conserve resources, particularly water. A direct economic signal is offered by an effluent charge based on the nature and volume of releases to the environment. The level will be based on the cost of treatment, and the flow discharged, in order to provide an incentive to set up treatment plants. The scope of the charges will also be extended to emissions and solid wasters. Charges provided a continuing incentive towards optimal releases.

These instruments will also have a distributive effect as the revenues will be used for enforcement, collective treatment facilities, research and promoting new investment.

The precise choice of economic instruments adopted will be determined by the ease with which releases can be measured, as well as prospective changes in technology and market structures. To deal with the range of pollution problems, a mix of regulatory and economic measures will be adopted.

The National policy is "the commitment on abatement of pollution for preventing deterioration of the environment". The complexities were considerable given the number of industries organisations and government bodies involved.

Maximum use would be made of a mix of instruments in the form of legislations and regulation, fiscal incentives, voluntary agreements, educational programmes and information campaigns.

The emphasis would be on increased use of regulations and an increase in the development and application of financial incentives.

ENVIRONMENTAL EDUCATION

Recent times have witnessed a global concern for the environment. Environmental education has gained a new thrust. The increasing environmental degradation makes it imperative that environmental education is included in the curriculum. As in the case of any other subject curriculum designing for such education entails formulation of objectives, content, methodology and evaluation procedures.

The present environmental conditions—shrinking forests and wildlife, the greenhouse effect, floods, famine, pollution—are man-made. It is in the minds of men that environmental problems originate. By sensitising the minds to environmental problems, education can assist in keeping man's demands within environmentally sustainable limits and thus improve the quality of life for all.

Environmental education is an integrated process dealing with man's interrelationship with his natural and man-made surrounding, population growth, pollution, source allocation, depletion conservation technology and urban and rural planning. Study of factors influencing the ecosystems, mental and physical growth, living and working conditions, decaying cities and population pressures have also been added to the above.

Environmental education curriculum experts have given the long-range goal of such education as developing a citizenry that is knowledgeable about the biophysical environment and is associated problems and is aware of how to become effectively involved in working towards the development of more livable future and is motivated to do so. From this one may discern that environmental education involves a multidisciplinary, inter-disciplinary and problem solving approach. It also brings into focus the need for developing values and ethics related to man's environment. As a step towards making people environmentally literate, environment education aims at creating awareness, knowledge and positive attitudes, motivation for actively participating in environmental problems, and skills for identifying and solving them.

Understandably, the curriculum must have cognitive affective and skill components. Respect for all forms of life, including plants and animals, must be inculcated. The emphasis in early years should be on the affective domain.

Environmental education may include study of the environment. Children may be taught about care of the greenery and of flowers, protection of birds and their nests, protection, care and feeding of animals in nature, the need for afforestation, and so on. This involves both the skill and affective domains.

A separate subject 'ecology or nature conservation' may be offered. The international working group on environmental education in school curriculum suggests inclusion of area and location, the cosmos and atmosphere, landforms, soils, minerals, plants, animals, social organisation, economics, aesthetics, ethics and language as content. In addition, ecological changes in nature and protection of rare or imperilled plant or animal species may be included. At the secondary level, the emphasis has to be on the cognitive as well as the skill or behaviour domain.

Two essential features of environmental education methodology are problem solving and valuing. In developing the problem solving ability, experimental encounters will be of help. An example of such encounters as investigating air pollution or solid waste disposal. Valuing or examining one's behaviour by clarifying for oneself one's own purpose, belief, attitudes and other value indicators is important in view of the fact that environmentalists have concluded that attaching too much importance to materialism and comforts is the root cause of all environmental problems.

The learning strategy adopted must be learning by doing, field studies, observation, practical conservation work, small group projects, class discussions, field trips, visits and simulation games. To begin with, students may investigate their own environment.

Research studies have shown that teachers are highly effective when they participate in the learning projects as team members, guides and counsellers. There is no ideal teaching model to which all students will respond favourably. Any effective model should provide opportunities for students to explore their environment sensorially, physically and intellectually.

Evaluation is an important aspect of curriculum planning. For effective evaluation, measurable goals must be there. For objective evaluation, an environment awareness test may be employed.

It is time that educational planners took up environmental education in all seriousness as that alone is an effective antidote to environmental illnesses.

ENVIRONMENTAL PROBING

Environmental Studies (EVS) should be included in the curriculum of primary schools. Textbooks on EVS should prescribe teaching-learning activities on selected ecological themes covering plants, animals and people, water, air and minerals, buildings, factories and their products, transport systems, as also objects in the way.

What should be the ecological perspective while teaching the above? The children in the rural and urban areas should be able to probe the environment in three dimensions— ''about'', ''through'' and ''for''. This approach is essential to realise the objectives of environmental education.

Children should be taken out into the immediate environment for probing under the guidance of the teachers. It can be a fascinating experience for them. An example, children observe columns of long red ants moving up and down a tree. Why are they doing that? The children find out that they are coming out and going into their nest. How is the ant nest constructed? What is it made of? How are the leaf edges joined together? What is the bio-relationship between the particular type of ant and the type of tree? What would be the effect on the ecology of the place, if that tree is cut down?

There are many possibilities of bringing materials from the environment into the classroom for probing. Children may be helped to make critical observations of them and also simple experiments with them. Small fish from the neighbouring pond can be brought to the class and kept in glass bottles containing the same pond water. Children can observe the movements of the fish and calculate their breathing rate by observing and recording the movements of the lower jaw. Let one of the children put a few drops of soap solution into the water and record the breathing rate. What is the tolerance level of the fish in relation to the soap solution? Can the experimental observations be related to the pollution in the village pond?

To sensitise people to major eco-issues, songs written to fit into their ethos and culture, are some of the powerful media to carry the message to rural people. Puppetry can also be used effectively. Posters can be designed and printed in attractive colours portraying the struggles of the deprived and marginalised communities against the dominating and exploiting forces.

A child could be the best communicator at home, and could persevere till the goal is achieved. In one case children are taken to a road to investigate the air pollution caused by motor vehicles. Each child observed the exhaust pipe at the near of the vehicle. They had developed a 5-point rating scale (very much smoke, much smoke, just smoke, very little smoke, no smoke) and used it for recording the data. They spent 30 minutes. The data were tabulated in the class and the interpretations were done by the children themselves. The teacher was the facilitator.

One child's father was a driver in a rich family. The child, after going home, asked her father whether a lot of smoke emanated from his car. He replied he was not concerned about that and that his duty was just to drive. The child insisted that he must find out and tell her. After a few days, the father told his daughter that black smoke came out in abundance, but he had not made a complaint to the owner. The child began pressing her father to tell the owner and get the car checked so that the air pollution could be minimised until he did. Interestingly, the owner sent for the child and gave her a gift in appreciation of what she did.

The teacher could devise a number of such interesting experiments and projects to develop eco-awareness in children and get them involved in action in the immediate community for the protection and preservation of the environment.

PART II

10. Global Issues and Strategies (Emanating from Earth Summit Fallout and Followup)

A GLOBAL APPROACH TO TACKLE ECO-PROBLEMS

Only through a global alliance the burning environmental issues could be tackled. This is because the problems transcend boundaries and what is now done collectively will determine the future.

The environment links all Nations. The atmosphere and oceans interact to provide the world's climate. Many great river systems are shared among several States. Pollution knows no frontiers, as it moves with the currents of air or water. Sustainability within a nation often depends on international agreements to manage shared resources.

People can alter their behaviour when they see that it will make things better, and can work together when they need to. This strategy is aimed at change because values, economies and societies different from most that prevail today are needed if we are to care for the earth and build a better quality of life for all.

Humanity which existed as a part of nature, had no future unless nature and natural resources were conserved. It asserted that conservation cannot be achieved without development to alleviate the poverty and misery of hundred of millions of people.

Sustainable development is improving the quality of human life while living within the carrying capacity of supporting ecosystems. The principles to be adhered to by a "sustainable society," lay particular emphasis on safeguarding biological diversity as well. Biological diversity should be conserved by a combination of measures to safeguard species and genetic stocks, the establishment and maintenenace of protected areas, and wider strategies that combine economic possibilities and conservation over entire regions.

Human impact on the earth depends on the number of people and on how much energy and other resources each person uses or wastes. The maximum impact that the planet or any particular ecosystem can sustain is its carrying capacity. Carrying capacity for people can be expanded by technology, but usually at the cost of reducing biological diversity or ecological services. In any case it is not difinitely expandable. It is limited ultimately by the system's capacity to renew itself or safely to absorb wastes.

Energy efficiency and the preventioin of pollution from the burning of fossil fuels is essential.

Life on earth depends on water. Our use of water is creating a crisis for much of the world. Global water withdrawals are believed to have grown more than 35-fold during the past three centuries, and projected to increase by 30 to 35 per cent by 2000. Current patterns of fresh water use cannot be sustained if human populations reach to 10 billions by 2050. Many countries already suffer serious water shortage. Competition among water uses is growing and exceeding the capacity of institutions to manage it. Water diversion and retention are having increasingly severe impacts on ecosystems.

For saving water, rainwater should be used efficiently through appropriate land use planning, water harvesting, choice and combination of annual and perennial crops, and scheduling of crop planting so as to conserve surface and groundwater for the most important uses. Better forecasting of rains is also needed.

Greater efficiency in allocating water uses and in using water can be achieved by charging consumers the full cost of water used. Full cost includes the cost of building and operating the water supply systems: of losses in distribution; and of protecting forests, wetlands and other ecosystems required to regulate flow and maintain water quality. Providing a reasonable supply of domestic water free of charge would prevent this measure from harming the poor and disadvantaged members of the society.

The low-income countries should adopt resource-efficient and environment-friendly technologies to double the supply of family planning services; improve education; provide basic health care and sanitation; rehabilitate degraded environments; conserve natural diversity; sustain agricultural productivity; expand reforestation; and increase energy efficiency and develop renewable sources of energy.

A GLOBAL PARTNERSHIP ON ENVIRONMENT

We have to focus attention on patterns of development that cause stress to the environment, demographic pressures and much more.

There is the need to integrate conservation with development, and to minimise the ecological harm (and hence the social and economic harm). Unless ecological consideration influence the development process as much as do social and economic considerations, and unless there is also an explicit policy to achieve conservation objectives, the prospects of avoiding ecological harm and making the best of living natural resources are dim.

The above is true to a greater or lesser extent of all the projects round the world.

Global warming, acid rain, ozone layer depletion, and other environmental problems call for our urgent attention. The bases for sustainable development, arrest further degradation of the environment, repair damage already done where possible, and even reverse clamitous current trends are to be found. Also protection of the atmosphere (climate change, depletion of the ozone layer, transboundary air pollution) protection of land resources (combating deforestation, soil loss, desertification, and drought) conservation of biological diversities, protection of freshwater resources, protection of oceans, seas, and coastal areas, and the rational use and development of their living resources environmentally sound management of bio-technology and hazardous wastes (including toxic chemicals), prevention of illegal traffic in toxic products and wastes, improvement in the quality of life and human health, improvement in living and working conditions of the poor by eradicating, poverty and stopping environmental degradation — all these are to be considered.

The underlying patterns of development which cause stress to the environment, poverty in developing countries, levels of economic growth, sustainable patterns of consumption, demographic pressures, and the impact of the international economy focus our attention.

Principles must govern the economic and environmental behaviour of peoples and nations, to ensure "our common future".

A blueprint for action in all major areas affecting and relationship between the environment and the economy, focusing on the period up to the year 2000 and extending into the 21st century should be drawn out.

Making available to developing countries the additional financial resources and environmentally sound technologies they require to participate fully in global environmental cooperation and to integrate environmental considerations into development policies and practices should be explored. These should include :

— agreement on strengthening institutions in order to implement these measures.

— conventions on climate changes, biological diversity, and forestry.

Action plan should cover every environmental concern from global warming to ozone layer depletion to rainforest protection to toxic wastes. Pollution and overconsumption of resources compel us to take environment more seriously.

The blitz on the vast rainforests of Sarawak, in Borneo, once the most splendid of their kind in south-east Asia, is predicted to consume them totally by the end of the century. It is making the loggers and their patrons and subsidiaries very rich while utterly destroying the lives and cultures of the once-thriving tribal communities of those forests. It is however, defended as ''economic progress'', a money-earner from the hardwoods sold primarily to Japan for making chopsticks, to be thrown away after use.

Such situations are exacerbated by the lack of an acceptable means of measuring conservation performance, making it difficult if not impossible to relate conservation goals to other readily measurable goals of activities such as agricultural production or reducing unemployment. More than 'a rule of law', a positive influence on development might be found in a policy of persuasion on the rich nations to constrain and reduce their demands on resources and move some part, not necessarily enormous, of their wealth to the poor ones, and on the poor ones to build in conservation, not all at once, into their decision-making. The third World countries cannot be forced to enact reforms to which they do not freely subscribe if we do not want to make them dig their heels in. Only real economic growth can create the capacity to solve environmental problems.

CHLOROFLUOROCARBONS AND ATMOSPHERIC POLLUTANTS/HAZARDS

The Earth's atmosphere consists of a mixture of gases which are under constant bombardment by solar radiation chiefly in the visible wavelengths. The most abundant gases such as nitrogen, oxygen and argon are all transparent to the visible radiation and also to most of energetic ultraviolet (UV) rays which are less abundant.

Energy Blanket

The survival of each component gas is governed by its chemical behaviour under such bombardment of radiation — its photo-chemical properties. Ultraviolet radiation shorter than about 240 nanometres (nm) is absorbed by oxygen molecule(s) which then photo-dissociate(s) into two oxygen atoms. These unite with other oxygen molecules to form ozone. In the process, surplus energy of the nascent ozone molecule is often transferred to nearby molecules as kinetic energy which slightly raises the surrounding atmosphere's temperature. (Nascent pertains to the atom or single compound at the moment of its liberation from a chemical combination). This transport of energy has been going on ever since oxygen became a permanent member of the atmosphere, millions of years ago and has made ozone play a crucial role in the development of life on earth.

Ozone accounts for only about three parts in ten millions of Earth's atmosphere but plays several critical roles in the radiation balance of the planet. While over 90 per cent of the ozone is in the stratosphere, peak concentrations occur 25 km above the equator and 15 km over the Antarctica. It is difficult to comprehend life on Earth without the protective layers of the atmosphere, the only life as we know it.

Ozone too is effective in absorbing particular short UV radiation (in the range 210-293 nm), releasing atomic oxygen. The fate of these atoms is similar to those that were photodissociated — transfer of kinetic energy, union with molecular oxygen, and release of fresh ozone. Interestingly, the added heating up of the surrounding atmosphere causes a temperature inversion between 15 and 50 km altitude from 58^0 to 2^0C. This defines the stratosphere as having temperature gradient while the temperature from the lower layer, the troposphere, increases steadily earthwards to an average surface temperature of 15^0C.

Natural Radiation Trap

The main inputs of radiation into Earth's atmosphere are the incoming solar radiation and the outgoing radiation emitted by Earth. The infrared (IR) wavelengths include only a minor component of the incoming solar flux, but carry all of the outgoing terrestrial radiation. Now, the absorption of IR radiation by a molecule hinges upon the matching of the molecule energy drawn from its internal vibrations and that carried by the radiation it intercepts. If the energies match, absorption is strong if not, no absorption occurs. Molecules with three or more atoms exhibit more internal vibrations than diatomic molecules such as nitrogen or oxygen. Therefore, atmospheric absorption is dominated by greenhouse gases such as ozone (O_3), carbon dioxide (CO_2), water particles (H_2O) and trace gases.

There should be a balance between the total amount of solar energy entering Earth's atmosphere and the outgoing energy from Earth. The absorption of infrared heat generated by the Earth's surface warms the lower layers of the atmosphere the troposphere. Because the air in the troposphere is warm, it radiates heat at infrared wavelengths back towards Earth's surface. The failure of this IR radiation to escape forces a higher level of terrestrial IR emission and keeps the surface warmer than it should be. This mainfests in the form of the greenhouse effect, a perfectly natural phenomenon and responsible for our survival.

Nevertheless, modern man's industrial activities are increasing the amount of trapped radiation leading to a rapid rise in global temperature by letting ozone-eaters such as the CFCs into the atmosphere. Every molecule of either of the two common CFCs — CCL_3F and CC_2F_2 has the same global warming effect as 10,000 carbon dioxide molecules.

The question of chlorofluorocarbons as a threat to the ozone layer first appeared in 1974 with the publication of a brief paper in the journal Nature. The study of CFC stratospheric ozone system (ozone in the upper layers of the atmosphere, 20-50 km high, the first line of defence against incoming ultraviolet rays) focusses fundamentally on CFC molecules which can hang on in the atmosphere for decades. They float up into the stratosphere by convective mixing, a process which refers to atmospheric motions that are predominantly vertical, resulting in vertical transport and mixing of atmospheric properties. Around 40 km altitude, they succumb to high-energy short wavelength radiation (such as UV) and initiate an 'ozone cleaving' process.

The UV radiation does not penetrate the lower atmospheres as they are absorbed by either ozone or molecular oxygen. However, after the CFCs are hit in the upper atmosphere, they split to release atomic chlorine (Cl). This then triggers a lengthy Clo reaction chain forming various oxides. This chain steadily eats away the oxygen atoms which would have otherwise refreshed ozone. Lack of ozone lets in more UV radiation leading to increased terrestrial reflection of infrared heat. This warms up the troposphere to uncomfortable levels and consequently raises surface temperatures — the global warming not to mention skin cancer.

UV radiation — the narrow spectral band which is thought to cause most biological damage appears to trigger two quite distinct immunological effects. One is confined to patches of skin that are actually irradiated while the other develops in the immune system as a whole.

The three kinds of skin cancer — basal cell carcinoma, squamous cell carcinoma and melanoma are rapidly climbing the list of human diseases. Langerhans cells in the epiderms of human skin are key players in immune surveillance. UV radiation gets them first, breaking down the defence in the skin. The pigment cells, melanocytes, produce melanin as a shield to absorb the damaging UV radiation. But most fair skinned people do not produce enough melanin to protect them from excessive exposure to sunlight. Sun bathing on the beach may therefore be not a pleasure to anyone. Then now comes with poison. Sunlight causes blood vessels near the skin's surface to carry more blood, making the skin hot, swollen and red-sunburnt.

Most epidermiologists concur on the causal relation between UV and non-melanoma skin cancers. Studies have shown correlation between these cancers and latitude, and by implication UV radiation levels. The closer a fair skinned person lives to the equator the more likely he or she is to get non-melanoma cancer.

The epidermiology of melanoma cancer cells is quite different. Curiously, the people most at risk are not those outdoors but office workers. It is suggested that melanoma is caused by intermittently exposing the body to relatively high doses of UV radiation and is often associated with burning sensation. This is underscored by the fact that melanoma is one of the handful of cancers that defy traditional links between poverty and disease. For most of these, particularly leukaemia and breast cancer, the reasons are obscure. But not so for melanoma: affluence and holidaying in the tropics go hand in hand.

CFCs-PHASING OUT

Phasing out of ozone destroying chlorofluorocarbons (CFCs) used extensively in the refrigeration industry and in air-conditioning has to begin.

Globally, action on controlling the use of CFCs had become urgent after the discovery of an ozone hole in the sky over the South Pole. During the Antarctic spring the hole became as large as the American continent, and more recently NASA scientists discovered yet another ozone hole over the populated areas of Canada and the U.S. These discoveries raised an alarm that incidence of skin cancer from exposure to ultraviolet radiation would increase and that damage to eyes, which were very sensitive to radiation, could be expected.

Scientists also predicted damage to crops as a result of excessive exposure to ultraviolet radiation and in South America recently there were reports of sheep having become blind.

A rough estimate showed that India consumed about 6000 tonnes of CFCs annually. Although the country had an installed capacity of about 9000 tonnes, it acutally produce half that amount. As against this the consumption in the U.S. was estimated to be 12 million tonnes a year and the countries of the North accounted for over 80 per cent of the total global CFC consumption.

The elements of the life-support systems include land, water and biological resources — all of which by themselves and by the inter-linkages between them are crucial to an agricultural economy like that of India.

CHLORINE IN STRATOSPHERE

The key problem in ozone destruction is not chlorine as such in the atmosphere, but the question is how the chlorine in chlorinated water would get into the stratosphere.

Chlorination to kill organisms in drinking water uses chemical forms related to pure molecular chlorine, Cl_2, a green gas. If it is given off into he atmosphere, it would ordinarily be decomposed by sunlight within hours, not leaving it enough time to reach the stratosphere. The related chlorine compounds are all easily removed by rainfall.

Some of the chlorine does react with organic materials, but it produced organic compounds, like methyl chloride, that mostly have lifetimes in the atmosphere of a year or so. So only a very small fraction would get to the stratosphere.

Methyl chloride is in fact a natural component of the atmosphere, in far larger quantities than would come from the chlorination of water, and the ozone layer has long been adjusted to its presence.

Chlorofluorocarbons, on the other hand, have serveral characteristics that ideally suit them to destroying ozone in the stratosphere.

First, they are not soluble in water, so they cannot be removed by rain. Second, they are transparent, and unlike the highly reactive green molecular chlorine gas, they cannot be transformed by available sunlight. And third, they are completely unreactive toward oxidising agents in the atmosphere.

ELIMINATION OF OZONE-DESTROYING CHEMICALS

Ozone in the upper atmosphere helps shield the earth from the destructive effects of ultraviolet radiation. Excess exposure to ultraviolet rays is known to cause skin cancer and cataracts and to suppress the human immune system. The natural radiation can also damage plants and ocean plankton.

Under terms of an international agreement called the Montreal Protocol, the U.S. and other industrialised nations had pledged to halt production of CFCs by 2000.

Ozone layer is being depleted at a much faster rate than had been believed.

The National Aeronautics and Space Administration reported discovery of "alarming" levels of ozone-destroying chemicals over Canada, U.S. and Europe.

POLLUTION OF OCEANS AND SEAS — NEED TO PREVENT

'The future of the planet is in the balance'. Of the seemingly unending environmental degradations taking place, pollution of the oceans is a major factor.

The oceans and the seas cover seven-tenths of the earth's surface, and no less than 98 per cent of that consists of sea water, with the remaining two per cent shared by freshwater lakes, rivers etc. Of the seas, an astonishing 35 per cent is claimed as "territorial waters" of nations. With the rapid growth of population and industrial technologies, both territorial waters and high seas have been abused in appalling fashion. There are many kinds of pollution to the seas: oil spills, industrial wastes, dredged spoils, sewage etc. Life itself began in the seas, more than three billion years ago. Ever since, the seas have retained their capacity to shelter and sustain the living world. Only in the second half of this century, an instant of geological time, have human activities threatened to disrupt and destroy those qualities. Our world is "a water world, a planet dominated by its covering mantle of ocean in which the continents are but transient instrusions of land above the surface of the all encircling sea". The changing shapes of the continents through time bear ample testimony to this statement. The continental drifts will continue indefinitely in the never-ending struggle between land and sea.

Before human action began to impact on the quality of the seas and sea environments, and their capacity to nourish life, change was dependent on natural factors: geography, interaction between land and sea, depth, warm and cold layers and currents, salinity and atmospheric events. These natural factors appeared to operate in such a way as to ensure, over measurable time spans, that the ocean's resources were inexhaustible or beyond damage. As human population and their capacity to apply growing technologies to the natural order began to escalate, the picture started to change all too quickly. Today, there is widespread perception that those resources are neither inexhaustible nor beyond injury: indeed, there is a mass of evidence that despoliation has advanced to perilous proportions in many areas.

Ocean incineration, or incinerating wastes in ships at sea, continues to the tune of 100,000 tonnes a year.

The North Sea is now one of the most polluted in the world. Each year 2.5 million tonnes of fish and seafoods are, or were, caught in it, about a quarter of all its seafood wealth. Each year, 2.5 million tonnes of solid and 1.7 tonnes of liquid wastes and 0.4 million tonne of oil and drill sludge are dumped into it. Of these, zinc accounts for 23,000 tonnes, lead 6,800 tonnes, copper 4,400 tonnes, chromium 4,200 tonnes, nickel 1,450 tonnes, arsenic 820 tonnes, cadmium 150 tonnes, and mercury 50 tonnes. A vital food source is being rapidly diminished and numerous health and environmental hazards being set on course. If this group of countries with requisite scientific knowledge, advanced technologies, sophisticated legal systems and highly vocal conservation bodies cannot act to save their own immediate marine environment, the prospect of countries much less geared for preventive action coming together cannot be rated high. The cultural complexities of their peoples from differing backgrounds and values must make an agreed approach exceedingly difficult. All this means that the future of the unprotected seas round the globe is at serious risk.

Today, the production and disposal of potentially hazardous wastes is part of the very structure of industrial consumer societies: societies which are not yet fully-fledged consumer societies are hunting willingly to that end and would feel cheated and deprived if they failed or were foiled in the attempt. So, the danger is going to increase multifold. The dramatic increase in pollution levels is to be traced to the unbridled use of petrochemical-derived synthetic industrial products. Examples are plastic containers and packaging, synthetic urea fertilizer, organochlorine pesticides, synthetic fibres, mercury for chlorine processing, and many more highly profitable but environmentally injurious inputs. Substitute products keep appearing and are promoted without really improving production processes to eliminate toxic byproducts: radical new methods being considered too costly. Wastes are disposed of less ostentatiously, in "economically sustainable ways," by burning, burying or dumping into smaller waterways.

Most of the industrial production around the North Sea is linked to two processes: fractionation, or 'cracking', of crude oil to produce a range of materials and solvents for manufacturing processes, and chlor-alkali electrolysis. The latter accounts for three-fifths of all industrial chemical production worldwide, and involve the breakdown of common salt (sodium chloride) into two feed items for the chemical industry: caustic soda (sodium hydroxide) and chlorine. Used jointly with solvents derived mainly from oil, these form the basis of chemical manufacture: plastics, agricultural chemicals, pharmaceuticals, industrial additives, cleaning products, paints and glues. Most of these cause major environmental, safety and health problems in manufacture, use and disposal.

Sea and marine environment disasters in the North Sea and its surrounds make a long list. The poisoning of the Baltic Sea by mercury, DDT and PCBs (polychlorinated biphenyls), and excess nutrients from agricultural runoff — the combined result is a sterile sea and closed (beaches); the pouring of toxic wastes into the North Sea from the 'Cheminova' chemical plants in Denmark for over 20 years (a hundred different poisons have been found in the marine environment, and a thriving fishing industry was totally destroyed); the drainage of dissolved poisons from the Rhine, Elbe and other rivers whose outlest straddle the 'Waddenzee' (a single copper works, the 'Nord-Deutscher Raffinerie,' had a hundred effluent discharge points passing huge amounts of copper, arsenic, cadmium and mercury into the Elbe, which was already scourged from massive toxic wastes from uncontrolled industrial operations in Czechoslavakia and the former East Germany); the discharge of sewage by pipes from Britain into the North Sea, the only North Sea country to do this still; and there are many, many more.

The fishing industry has been one of the prime food producers for man. Over the last half-century, overfishing to the extent of some 400 per cent over previous catches has denuded global fish stocks, and the stage has been passed where supply can be expected to match demand. While this sheer excess fishing has been the main cause for stock depletion, pollution's contribution has become increasingly decisive.

Pollution has two effects. It reduces yields of seafoods depending on the type and amount of pollutant, and it makes seafoods unfit for human consumption, or for that matter consumption by other lifeforms. The first of these is obvious: by killing off, yields decrease. The second is more difficult to detect, and has been the subject of much prevarication.

There are numerous examples of fish and seafood population destruction and, depletion from pollution round the globe. These begin with the poisoning and killing of small inshore marine life (which would normally thrive from the nutrients that rivers bring from the highlands) which is the very subsistence of fish, other sea creatures, and sea birds. In Britain, the 'Humber' estuary supported a thriving fishing industry. It is dead now, from the poisoning of the estuary by discharges from factories that now line the river. Fish stocks in Chesapeake Bay' off Maryland and Virginia in eastern U.S. have been all but destroyed by effluents from some 3,000 factories that had become accustomed to using the Bay as their waste dump. The Mediterranean Sea is in an advanced stage of degradation, and continues to receive the raw sewage from three-fourths of the population in the countries bordering it, besides hosting massive amounts of oil waste, heavy metal residues, and agricultural runoffs. The Caribbean Sea, home to industrial and agricultural effluents from the Mississipi river basin and the Latin American countries around it, is heavily contaminated. And the most sewage-polluted seas of all are the shore areas of densely populated south Asia. Marine animal and bird populations in each of these seas have been decimated: for instance, 14,000 common seals died in the Baltic and North Seas in 1988 from pollution and viral epidemics.

The effects of contamination on marine animals and the birds which feed on them, including genetic changes and reproductive failure, and dangerous. The pernicious results on human consumers of contaminated seafood is dramatic and measurable. The Japanese experience of 'Minamata' disease from mercury-tainted fish and shellfish landed at many Japanese ports, and the ongoing serious occurrence of 'hepatitis A, Polio' and dysentery along the Mediterranean shores, are cases in point.

A quarter of the world's protein is supplied by fish and other sea-foods, and the fishing industry provides a livelihood to millions of people round the world. A real advance should be made to protect the marine environment.

ENVIRONMENT IN HEALTH AND DISEASE

The role of environment in health and disease is important.

Health of an organism whether of plant or animal species to a great deal depends on the environment in which it is born and brought up. In fact the survival, well being and productivity of a human being is very closely linked with the environment.

Right from conception onwards throughout the long journey from womb till tomb, environment has a major role in the well being or otherwise of an individual. The environmental calamities affecting the health of thousands resulting in death and disability due to "Killer smog" and mercury poisoning from seafoods that occurrred in some countries are still fresh in our memories.

With the modern fast communication system, environment related health hazards cannot remain restricted to a few local pockets and hence any strategies for the control of such problems have to be planned at a global level.

The definition of health itself as per the World Health Organisation, is a state of complete physical, mental and social well being and not merely the absence of disease and infirmity.

Pollution has no intercountry boundaries or international borders. Pollutants emitted from the vehicular exhausts, thermal power plants or the chemical factories and poured into the atmosphere or the rivers are carried to far off places wherever the direction of the wind and water takes them. It is difficult to estimate at present what will be the biological effects of the radioactive fallouts from various nuclear tests and accidental nuclear leakages on human and plant life, but a higher occurrence of leukemias, and other cancers could be the fear for the inhabitants staying near these areas.

Environment can be both a cause as well as cure of many diseases. While fresh air, pure water, and mother earth are the natural positive health environmental factors, we have on the other hand some man-made disease producing environmental factors such as smoke from chimneys, cigarettes or vehicles polluting the air we breath and industrial toxic waste products contaminating the water we drink and also the contaminated earth in turn gives us toxic vegetables and fruits that we eat.

Aberrations at the psychic level are the cause of many disorders called psychosomatic diseases such as bronchial asthma, some skin diseases, abnormalities of heart rhythm (arrhythmias), hypertension, angina and irritable bowel syndrome. Incidence of diabetes mellitus, hypertension and coronary heart disease has been found to be higher in certain population after they have moved out to newer surroundings thereby once again pointing out the powerful influence of environment on disease patterns. Higher occurrence of diabetes and coronary heart disease among Indians in U.K. and of hypertension among black migrants in the U.S. and ischaemic heart disease among the American Japanese are some of the examples of the effects of environment on health and disease.

The internal environment has a strong role to play in the causation and aggravation of cardiovascular problems including hypertension, angina, myocardial infraction, arrhythmias and cardiac failure. Emotions and stressful situations can create a hostile environment which may prove harmful for the cardiovascular system.

Emotions, may be divided into (a) positive (b) negative.

Stress and tension are often mentioned as risk factors for many diseases including some cardiac disorders and hypertension. It is very difficult to define stress which is essential in life as a drive to work and is, therefore, needed for a productive and useful living, useful for self and for society.

The factories and thermal power plants which on the one hand give us electricity and numerous articles for our living comforts also give us smoke for atmospheric pollution and poisons for water and land pollution. Lung diseases like chronic bronchitis, emphysema and chronic cor polmonale are the result of air pollution, while water pollution is responsible for several skin, gastro-intestinal and liver disease. Industrial and vehicular noise pollution causes many hearing and cognitive disorders. Rain water at one time was considered of great therapeutic value for washing clean both the plants and animals but now we fear the "acid rain" due to pollution of atmosphere with various chemicals and toxic material as a result of the sulphur coming from the automobile exhausts and thermal plants.

In the context of health and disease, environment may be divided under two headings viz. (i) External and (ii) Internal (milieu interior). Factors like air, water, earth and its products, various food articles that we eat and friends and foes that we have around us comprise the external environment.

The blood with its various components that circulates in our body, hormones, our emotions, and the state of our mind constitute what we call internal environment. There is a close association and interdependence between external and internal environments.

Many diseases are not so much due to the gift of genes but the result of imbalances in the internal and external environments. Diseases like high blood pressure (primary hypertension), heart attacks (coronary or ischaemic heart disease), over-weight (obesity), certain abdominal diseases like peptic ulcer and irritable bowel syndrome, are some of the problems which are largely the result of faulty environment. Partly external and partly internal. Psyche or mind has a powerful role on the brain, heart and organ systems of the human body.

In the individual a certain stress is a drive for useful living while in another a similar stressful situation becomes a 'distress' for him and is inimical to his life. The definition of tension or stress thus is a relative one depending upon the circumstances and training of the individual.

The nearest, one can define stress is: "A situation resulting from an imbalance between the environmental demands and the individual's capability to meet these demands". Here again when talking of stress the important component of definition of stress is related to environment.

Stressful situations and tense or hostile environment are often held responsible as a major risk factor for cardiovascular disability and sudden cardiac death. Environmental factors thus play a great role in causation and/or aggravation of several diseases in general and of the cardiovascular system in particular.

Let us now look at the curative role of environment and its role in preventing disease and preserving and maintaining positive health. It is a well known fact that many of the modern diseases like diabetes, hypertension and coronary heart disease are less prevalent in rural populations than in urban dwellers. The same individuals when they migrate from villages to cities or from one country to another, after a length of stay tend to show higher incidence of these diseases. Both external and internal environments change. The internal environment depends to a large extent upon the external environmental factors being peaceful or hostile.

The clean unpolluted air, away from the chimneys and heavy vehicular traffic of industrial townships and large cities, the morning fresh air, free of fumes and smoke has a great cleansing and purifying effect on the lungs and the body as a whole. Health and longevity of populations residing in mountaneous areas is in general better than those living in the plains. This again is attributed to unpolluted air, clean drinking water and enough physical exercise.

Environment has an important dual role in health viz. in causation and cure of disease and in preserving and maintaining positive health.

With the global warming due to environmental pollution, melting of the icebergs, rise in the ocean level and submerging of the land, and whatever grows from the land if that too is going to be polluted from the industrial toxins and fumes from surface and air transport, we will have to seriously study the "environment" for fruitful health care of the Nations.

Environment Strategy, Safeguarding the Environment

The new environment policy relates to pollution abatement and 'clean (non-polluting) technologies' rather than 'clean up technologies'. It is known that over 50 per cent of the pollution load is accounted for by industrial waste, although in terms of volume, this source contributes to a much smaller proportion, and that a chunk of it comes from the major and medium sectors. In spite of the multifarious anti-pollution laws, rules and guidelines as also the institutional framework that have emerged in recent years, the problem remains as staggering as ever for a variety of reasons, the most obvious of them being of course that it is unlikely that violations would meet with swift and stern retribution.

But then, inasmuch as pollution like every other malady is best tackled at its root, it is only right that the factories and other enterprises concerned are required to make a self-assessment of their role in this critical area. As a prelude to the statutory stipulation of environmental audit all manufacturing units should come out with statements every year on the nature and extent of the damage they inflict on the eco-system. This way they are made to realise their accountability to the community.

The want of authentic and complete data about the chemicals in use either as end-products or as inputs in manufacturing processes has been a major handicap to any meaningful evaluation of environmental impact. With the number of such chemicals increasing year after year, the problem gets compounded and this is something even advanced countries are finding difficult to cope with. In this context, mandating the companies to publish status reports about the effects and emissions should in itself be of great advantage.

No less significant is the idea of using fiscal instruments for achieving the pollution abatement goal by inducing a shift from the curative to the preventive approach and internalising the costs of pollution and resource conservation. A sectorwise survey covering such fields as cement, sugar, distilleries and pulp and paper put the capital and recurring costs of effluent treatment facilities (as stipulated by law) at levels far beyond the capacity of even the big industrial houses.

TO HAVE CLEANER AIR

National limits should be established for the amount of two types of chemicals, used mainly by semiconductor and automobile plants, that can be discharged into the atmosphere.

Exposure to the two chemicals, trichloroethylene and tetrachloroethylene, can cause liver and kidney damage. Tests have also shown that they can cause cancer.

Although there are limited discharging the chemicals into water there are no national regulations covering their release into the air. Some cities, however, have established their own limits on the chemicals in the atmosphere.

The chemicals cause greater damage when they are inhaled than when they enter the body through drinking water.

Trichloroethylene is used as a washing agent for metals and a solvent for various types of chemicals. Tetrachloroethylene is mainly used as a washing agent for metal and in chlorofluorocarbons (CFCs) which are used in airconditioners and some types of plastic foam.

A study conducted found that in highly polluted cities, the residents are absorbing particularly high levels of the two chemicals. The average adult absorbs 101 micrograms of trichloroethylene and 124 micrograms of tetrachloroethylene a day through the air breathed. By contrast, the average for all other residents was only 8.2 micrograms of trichloroethylene and 21 micrograms of tetrachloroethylene.

Air pollution from the two chemicals should be reudced.

CLEAN ENERGY PROJECTS

Germany has been in the forefront of global efforts to create, expedite and streamline programmes in various spheres of sustainable development, its orientation in the area of energy supplies and their optimal expansion in developing countries through both technical and financial assistance.

In tune with the precept of ensuring environmentally-sound use of biotechnology, Germany is active in the promotion of the indisputably inexpensive energy resource of biogas in developing countries. It has gathered enough evidence from its experience to substantiate its view that intensified technological co-operation under the purview of development promotion and anticipatory environmental policy were urgently required. While felling of trees on a wide scale for the purpose of firewood and its extension of destroying tropical forests is a familiar phenomenon, several countries scarcely have recourse to other energy sources.

It is here that the tried and tested renewable energy source of biogas with a high degree of utility under correct operating conditions has developed in direct proportion to the growing ecological consciousness in both industrial and developing countries around the world. A proper application of techniques for the utilisation of the biomass resulting from organic wastes of fruits and vegetables, residues from harvests, wastes from animals and humans as well as industrial refuse has now resulted in the shaping of a biogas technology.

Experience

A decade and more of experience with biogas projects in the developing world has testified to the efficacy of this technology that could be applied with equal success by industrial and developing countries alike. Indeed, what began as an experiment in the early Seventies in the developing countries has now graduated the small biogas plants in rural areas to the installation of large plants that also turn out organic fertilizers, among others.

India has been prominent among those seeking to expand their energy sectors in the non-conventional area, with some 1.2 million biogas units now operating. The low costs incurred in the setting up of small biogas plants and the pollution-free nature of the systems used have made this technology a prime candidate for private enterprise, aided by the State to a certain extent in promotion. Development aid has not only come in with strong support for multiplying projects in the area but also has played a role in the propagation of this philosophy that at once answers the large scale needs of the rural sector in consonance with the crying needs of the environment.

The efforts are for increased energy efficiency and its rational use, promotion of new and renewable energy resources, and the inclusion of industry, households, transport and agriculture. Risk assessment and compliance with safety standards are an integral part of biotechnology application as well as on "development of internationally binding safety directives for the application of biotechnology." The role of biogas, especially in rural areas with abundant natural resources, is unquestionable. Biogas has no harmful emissions and is burnt without any health hazard or environmental pollution.

Rural Economy

The rationale for a large-scale adoption of biogas technology in countries blessed with adequate local resources and entrepreneurial dynamism, is self-evident. So is its relevance for the environment in the context of development. New impulses are expected for a broadening of the energy source which has an important contribution to make towards easing of energy problems in rural households of developing countries.

New concepts have emerged for enlarging the uses of biogas refining its applications. Methods for highlighting the use of biogas in their finanical and socio-economic aspects as well as their evaluation and monitoring have also come a long way. Improving the access, in particular of developing countries, to environmentally compatible technology and making new and additional funds available for measures taken by developing countries in the field of national and international environmental protection are urgently called for.

Norms for Environmental Clearance of Projects

The following are the norms for clearance of various new projects and expansion or modernisation of existing industry.

Considerable adverse environment impact has been caused due to degradation of the environment with excessive soil erosion and water and air pollution on account of certain development activities, thereby engendering not only the destruction of natural resources like forests, mangroves, wetlands, rivers, lakes, genepool reserves and vegetation cover which is fast dwindling in large parts of the country, but also affecting the health and very survival of living being — both animal and human.

As it is necessary to protect and improve the quality of environment by controlling pollution of air, water and soil along with biotic pressure on natural resources, which is so intense, that natural biological and genetic wealth is threatened with severe damage.

Since certain development projects should be carried on within the carrying capacity of the eco-system through judicious location beyond defined distances from eco-systems which will otherwise come under stress, so as to ensure that developmental activity takes place in harmony with the environment and improvement thereof.

The aforesaid goals can be achieved only by careful assessment of a project proposed to be located in area, on the basis of an environmental impact assessment of each project and the necessary environment management plan for the prevention, elimination or mitigation of the adverse impacts, right from the inception stage of the project. The procedure for seeking environment clearance of projects — (1) Any person who desires to undertake any project in any part or the expansion or modernisation of any existing industry or project shall submit an application where environmental clearance is required.

In case of the following site-specified projects, mining, pit-head thermal power stations, hydroelectric power projects; and multipurpose river valley projects, and a preliminary site clearance will be required from the concerned Authorities before initiating any investigations involving cutting of trees, drilling, digging or construction of any sort temporary or permanent.

The said site clearance will be granted for the sanctioned capacity and will be valid for a period of five years for commencing the construction. The detailed project report submitted with the application shall be evaluated and assessed by the impact Assessment Agency in consultation with a committee of experts.

The said committee of experts will have full right of entry and inspection of the site, or as the case may be factory premises at any time prior to, during, or after the commencement of the operations relating to the project.

The Impact Assessment Agency will prepare a set of recommendations based on technical assessment of documents and data furnished by the project authorities supplemented by data collected during visits to the site of factory, interaction with affected population and environmental groups. The recommendations and the conditions subject to which environmental clearance is given may be made available to parties concerned.

In order to enable the Impact Assessment Agency concerned to monitor the effective implementation of the recommendations and conditions subject to which the environmental clearance has been given the project authorities concerned shall submit a half-yearly report to the agency concerned.

Schedule I

List of projects requiring environmental clearance from the Central Government.

Atomic power, thermal power, multi-purpose river valley projects, ports, harbours and airports, railway lines (involving acquisition of non-railway land), refineries, fertilizers, pesticides and insecticides, petrochemicals, explosives and accessories, drugs and pharmaceuticals (except formulations), production of plastics, rubber-synthetic, asbestos and asbestos products, sodium or potassium cyanide, primary metallurgical industries (zinc, lead, aluminium and steel), integrated steel plants, tyres/tubes or tracks, jeeps cars and other heavy vechicles, alkalis (NaOH) integrated paint complex, man-made fibres (synthetics, rayon, nylon and polyester). storage batteries with lead processing, incineration plants for hazardous waste and chlorinated hydrocarbons and all projects with threshold criteria above those specified in Schedule II.

Schedule II

List of projects requiring environmental clearance from the State Government.

Ceramics, coal washery, briquetting, carbonising plant, engineering (tubings, castings, rolling mills), refractories, pipelines (RCC, steel, seamless), calcium carbide, carbon black, pharmaceuticals (formulations), electroplating, storage batteries (non-lead processing), alkalis (Na_2CO_3 and $CaCO_3$), plastics processing, tyres/tubes of two-wheelers and cycle rickshaws, retreading of all types of tyres, paints (varnish, coal tar chemicals, pigments), dye single industry, soaps and detergents, food processing (meat, fish, animal products), milk processing, paper products, incineration plants for non-hazardous waste.

Hydro-electric, new projects up to 10 MW, all powerhouses up to 10 MW to be located on existing canal falls or existing dams with already embeded penstocks, cement up to 200 tpa, steel plants up to 50,000 tpa, leather tannery up to 5,000 skins PD, distilleries up to 150 kld, sugar up to 4,000 tpd cane, textile up to 500 mts/d, pulp, paper and newsprint up to 33,000 tpa, dye intermediates up to 5 tpd, thermosets phenol and urea formaldehyde up to 5,000 tpd, acids up to 500 tpd, vegetable oil processing up to 500 tpd, foundaries up to Rs. 20 crores. Capital cost communication up to Rs. 20 crores. Tourism and other projects written 1 km of HTL of sea, all locations with an elevation of more than 1000 mts, up to Rs. 5 crores, irrigation up to 2,000 hectare command, mining up to 5 hectare lease, roads (in Himalayas or involving forestland) up to 5 mm length, industrial estates up to 100 units and industrial township (new & expansion) up to 5000 dwelling units.

BIOMASS AS A FUEL FOR ENVIRONMENTAL BENEFIT

The primary energy sources like coal, oil and natural gas are rapidly depleting, which by the end of the next century, will further increase our dependence on renewable resources like solar and biomass. The Survey of Indian Industry, 1990, reported that coal contributes to over half of the commercial energy needs, mainly the coal hungry thermal power stations and steel plants. Environmentalists have more than often warned about the serious implications of burning fossil fuels — the greenhouse effect, acid rains (due to CO_2 and SO_2 emissions) and the ozone layer depletion.

The biomass is one renewable resource with more environmental benefits than problems. An important benefit is the use of wastes that would otherwise be a pollution source. Energy from biomass comes in the form of solid, liquid and gaseous fuels with appropriate conversion technologies. By the turn of this century, up to ten quads per year of oil and natural gas can be displaced by wood and other biomass. However, the actual displacement would depend on the conversion methods chosen, not to mention the market for the alternative fuel.

Gasification and conversion to methanol seem to offer great promise towards displacement of conventional energy sources. In fact, gasification, the more energy efficient of the two conversion technologies, can serve as a direct substitute for oil and natural gas. Methanol can directly replace petroleum fuels, although conversion of biomass to it is less efficient than gasification or direct combustion.

Some important factors that could determine the amount of conventional fuel, in particular oil and natural gas, that could be displaced by biomass are management of available crop land, how the biomass is converted to useful work, fuel and heat, and what the converted fuel is used for.

Fuel Cycles

Four fuel cycles which can contribute a significant amount of energy are :

— Wood for direct combustion, gasification and alcohol production;
— grain and sugarcane crops for alcohol;
— agricultural and crop residues for alcohol;
— animal manure, aquatic biomass, industrial and municipal liquid effluents for biogas through anaerobic digestion.

Direct comhustion is limited by existing technology to applications such as boilers and water heating. Combustion of solid biomass can join coal and solar energy in displacement of oil and natural gas. If used primarily in the domestic sector, it may be technically possible for biomass combustion to substitute conventional fuel to a large extent.

The uses of crop land to produce alcohol fuel feedstocks vary in the efficacy at displacing conventional fuels depending on the yield per hectare. Ethanol from sugarcane molasses and starchy materials like casava, offers a good net premium fuel displacement. Wood, agricultural and other lignocellulose residues represent a major proportion of the total biomass available.

Use of Fuel Feedstocks

The three methods of using these fuel feedstocks are direct

combustion, gasification and liquefication to methanol or ethanol. Gasification (carbonmonoxide and hydrogen as fuels) and direct combustion are the most energy efficient processes while liquid fuel combustion is the least. However, taking advantage of the octane boosting properties of alcohol in gasoline blends, the options can be made more comparable.

Of the three converted energy forms, alcohol is the most versatile while the least is direct combustion. Alcohol can be used in the trasportation sector and by all stationary fuel users. In the latter case, synthetic gas from biomass can be used, too. Alcohol fuelled automobiles can be around 20 per cent more efficient than their gasoline counterparts.

Gasification can be the largest potential substitute though limited to pit-head, on-site and other stationary uses. The principal constraint is the low heat content of air-blown gasification, which given the transportation costs limits its use to close coupled gasifiers.

In India, there is a vast potential for the conversion of animal wastes to gaseous fuel, namely, methane through anaerobic (without oxygen) digestion. In an agrarian country like ours, agriculture is integrated with cattle population in the villages. Often there is a shortage of conversional fuels for their domestic consumption. Thus the villages depend heavily on forest biomass and agricultural residues for their basic needs. Also, inefficient burning of biomass leads to serious degradation of the surrounding forests. The policy of burning agricultural residues is probably debatable as the soil fertilizer and animal feed cannot be dispensed with.

Moreover, biomethanation of animal manures and human excreta is an attractive concept through anaerobic digestion in view of its beneficial environmental management. Also it results in a fertilizer that can be used not only for the crops but also as animal feed.

Direct burning of cow dung, as done in the villages, is inefficient, besides being an environmental hazard and a loss of fertilizer. Each tonne of cow dung burnt may mean a loss of about 50 kg of food grain. An estimated 278 million tonnes of dung is produced annually by 257 million heads of cattle. Even if one half could be diverted for biogas generation, it would equal 80 million tonnes of coal.

Biomass as an alternative source of energy rests on a few issues. The first one is reliability of its supply. The consumer will want to be sure of a long term and steady supply of biomass energy resource. The second is the collection, transportation and storage costs. Here the density of the biomass and proximity of its source are important. The third issue is the scale of operation. A small scale unit with access to biomass may find the going attractive as the demand is limited and resource assured. For instance, an agrarian family and small community can run a biogas unit on the local cattle dung supply and night soil. The fourth is the renewability of the resource, even if it works out less economical than conventional fuels.

Gasification of biomass may be less complex and more economical for small scale operations than coal gasification. Moreover the dispersed nature of biomass generally limits its application to smaller operations than those for coal.

With regard to conversion to liquids and gases, it is important to continue efforts to develop new and more efficient methods to convert wood, animal manures, crop residues and other lignocellulosic matter to gas and alcohol. Although some processes are commercially ready (alcohol and biogas production, for instance), the realisation of the full biomass potential is likely to require additional developments in these areas.

Economic and social development and poverty eradication are the first and over-riding priorities of developing countries. The concern for environment cannot be allowed to be used as a pretext to force countries to take steps that would result in slowing down their development.

The funding mechanism shall operate "within a democratic and transparent system of governance". The Global Environment Facility, as it exists would have to be restructured. The biodiversity issue is to be linked to biotechnology.

The biodiversity is closely linked to the issue of intellectual property Rights. While the need is there to preserve biodiversity, any biotechnology that flows out of research as a result of access to biomaterials, must not come under the IPR regime. The South agrees to preservation of biodiversity and allow free access to it for research provided the North agrees to make technology flowing from it also accessible.

The North, while stressing on the need for effective steps to preserve biodiversity as a common heritage, would like free access to biomaterials but would like to apply an altogether different yardstick to any biotechnology that flows out of research on these materials. In their words, the North would like to use the genetic resources from biodiversity but would like to exploit commercially any technology flowing from it.

There is general need to cutback on carbon dioxide and other greenhouse gas emission to prevent a disastrous warming of the earth, and an agreement setting down deadlines for cutback on global fossil fuel consumption to reduce greenhouse gas emissions is necessary.

LOW POLLUTION BIOPOWER FROM BIOMASS

A clean approach to generating electricity from biomass is by drying whole trees and burning them at high temperatures.

The technique raises the efficiency of combustion and reduces pollution, it also avoids the cost of chipping the wood.

The dried wood is loaded onto a conveyor belt and fed into the generator which is modified to burn wood. The dry wood burns at a temperature of about 1500°C. The combustion efficiency is between 33 to 35 per cent, compared to about 25 per cent for moist wood.

Combustion of three levels — the wood itself, vapours about it, and charcoal below it — helps increase efficiency and reduce emissions of nitrogen oxides.

SOLID WASTE MANAGEMENT

Solid wastes are an index of our lifestyles; the wastes from most infra-human levels are biodegradable and hence recyclable. Reduction, reuse, rejection and recycling are the four strategies for solid waste management.

Solid wastes are more an urban and much more, a metropolitan problem. Collection and disposal are the two major concerns of solid waste management.

Home is the starting point for all municipal wastes. We have to monitor the reduction of wastes at the source, to reuse bottles and containers, to dispose off the recyclable paper, polythenes, plastics, metal and glass scrap etc., and to dump the rest neatly into street bins.

Wastes dumped into street bins are first taken care of by natural scavengers like pigs, dogs, rats and crows, which incidentally disseminate the contents and then by self-employed rag-pickers who help in recycling and in reducing the volume of wastes. However, the amount of wastes produced in our homes and the careless way in which it is strewn on to the streets, render it practically impossible for any municipality or corporation to cope despite the enormous number of sanitary workers and vehicles meant for transporting the wastes.

Land-filling of low-lying areas can reclaim sites for construction of housing colonies or for public buildings. However, rainwater that soaks and seeps through such dumps may pollute the water-table with highly toxic or pathogenic contaminants. Such dumps may emanate biogas or methane which is one of the greenhouse gases, but which could be collected as fuel gas. Landfills in the neighbourhood of residences are resented because of the stench they emit and the swarms of houseflies and mosquitoes they breed. Accelerating composting, by using earthworms on such community dumps, also called "vermi-composting" can be popularised.

Incineration of combustible wastes could be employed for generating energy from steam, or for lime and brick-kilns. Fuel pellets or brickets could be manufactured out of combustible wastes, if they have appreciable calorific value. Domestic or community incinerators are welcome but smell from dried leaves and certain other wastes add to the carbon dioxide levels, another greenhouse gas.

Recycling of wastes, converting them into wealth provides livelihood for some; recycling is best done in poorer developing countries. Polythene sachets and bags are not only non-biodegradable, but if thrown carelessly, can also hold enough water to help mosquitoes and houseflies to breed. Unless residents of each street are educated to help in the quickest and safest disposal of solid wastes, our streets and our cities would be uninhabitable.

NOVEL GARBAGE CLEARANCE

An integrated Waste Management scheme comprises setting up of a plant that converts garbage into fuel pellets and construction of pre-fabricated community toilet complexes in slum areas.

Construction of 50 garbage processing plants and about one lakh toilet complexes would cost about Rs. 300 crores. A pilot plant processes nearly 120 tonnes per day (TPD) of municipal waste to produce 80 TPD of Refuse Derived Fuel (RDF) pellets.

The combustible pellets which have a 10 per cent ash content (against 35 per cent in coal) and do not contain sulphur but only carbon dioxide, could well be a suitable alternative to coal for textile mills and chemical industries. They can be used as domestic fuel also.

Linked with the garbage processing plant, is the construction of a chain of community toilet blocks in the heart of slum areas, based on the pre-fabricated ferro cement technology for speedier mass replication.

ECO-STRATEGY

Conservation Strategy and Policy on Environment and Development, should seek to 'weave environmental considerations into the fabric of national life and of our development process'

Action should be taken through various regulatory and promotional measures, and development policies from an environment perspective in key sectors such as agriculture, energy generation, forestry, industrial development, transportation and human settlements.

International cooperation on global environmental issues should not slow down development in name of environment. The resources that the country has, are required urgently for education, nutrition, health services, drinking water, housing, sanitation and investments in industry and infrastructure for agriculture.

The development process in the short run can add to the emissions which cause global problems (such as carbon monoxide emissions) but these can easily be compensated by a marginal reduction in emissions from industrialised countries. While the country will strive to adopt environmentally sound policies, it is only with new and additional funding and with easier transfer of technology that we will be to strengthen its capacity to deal with environmental problems.

Population control strategies are crucial for any meaningful environment policy and the relationship between the life support systems and increasing population must be stabilised. The loss of forests and therefore wildlife habitats had endangered an estimated 1500 species. Gene erosion could seriously affect the prospects of getting full economic benefits from recent advances in molecular biology and genetic engineering.

Land erosion and water pollution, loss of grazing lands as cattle population increases, the threat of a food crisis in the future, unless land and water resources are better managed, air pollution and the need for making industry responsible in regard to anti-pollution are other issues to be tackled.

Environment brigades be set up for increasing environment awareness among the youth for reporting illegal acts pertaining to forests, wildlife and pollution control laws and for monitoring and collection of air, water qualities.

Global Warming Threat to Avert

By the end of the century, the average global temperature will be higher than ever in the past thousand years. And by the end of the next century this trend could make the earth 3 to 4°C hotter taking the world's climate back to what it was two or three million years ago, a time when lush green vegetation covered Siberia and northern Canada and dinosaurs roamed the earth, when there might not have been permanent polar ice-caps, and when the sea-level was a good one metre higher than it is today.

Why people should worry about the earth getting a bit hotter? Why does this expected climate change need international preventive action? Why do heads of over a hundred governments need to get together to deliberate this issue? And finally, why is this happening and what will be the implications for human beings?

Global warming of 3 to 4°C could lead to alarming and unpredictable changes in established weather patterns that rainfall and wind patterns could change dramatically and seriously affect agriculture.

Apart from this the rising sea-level as a result of thermal expansion of water (water when heated expands to occupy more space) and melting of the polar ice-caps would flood coastal areas, submerge small island-nations and destroy agriculture in coastal areas by rendering the land saline.

Today the climate change problem is very real. The eighties witnessed three or four of the hottest summers recorded and it is predicted that the trend would be confirmed by the end of this decade. Global warming may be already with us, with the average temperature now higher by about half a degree celsius, as predicted by the computer models and directly related to the increasing amount of carbon dioxide in the atmosphere.

The 'greenhouse effect' is caused by the increased amounts of carbon dioxide, water vapour and other trace gases in the air which trap the heat emitted after the sun warms the earth's surface during day. These gases build up in the atmosphere as a result of man's activities (mainly burning of fossil fuels which produces carbon dioxide as a by-product), trapping more heat. It is as if the blanket of air around the earth, which keeps the earth warm, gets thicker and thicker.

The expected increase in average global temperature by 3 to 4°C over a century may seem small, but that it is the difference required to melt the polar ice-caps and increase the sea-level by a metre. An increase of one degree in the average global temperature in effect is the same amount of extra heat over the year as a rise by 365 degrees in one day, more than enough to devastate agriculture.

A doubling of the carbon dioxide in the atmosphere will increase the average global temperature by nearly 2°C. Carbon-studies of trees suggest that around 1850, just before the industrial revolution took off, the carbon dioxide concentration in the atmosphere was 270 parts per million (PPM). By 1957 it went up to 315 PPM and today an estimated 5 billion tonnes of carbon or its equivalent is belched out annually by man's economic activities—apart from the carbon dioxide given out by the life process of breathing.

It is also estimated that if no action is taken to stabilise this emission of carbon dioxide and other greenhouse gases, it may add 20 billion tonnes of carbon—equivalent to the atmosphere, and the natural carbon cycle which has, over millions of years, maintained the balance of carbon dioxide in the atmosphere, could break down.

In such conditions life may continue to flourish but will man? That is the big question posed by the global warming problem.

Will man's ingenuity and new technology meet the challenge head-long? Will we see an 'energy revolution' through development of renewable and non-polluting solar and wind energy to meet the challenge just as Malthusian doomsday prophecies of death by hunger were given the lie to by the green revolution?

Statistics show that currently the richest countries with 20 per cent of the global population are responsible for 80 per cent of the carbon dioxide build-up in the atmosphere. If the total period of the industrial revolution is taken their contribution could be as high as 90 per cent or more.

The recent climate change convention, notes that "the largest share of historical and current global emissions of greenhouse gases has originated in developed countries". The "responsibility" for the damage to environment has been placed squarely on the shoulders of the developed countries. For if 'the polluter pays principle' is adopted and applied, then the First World owes a debt to the Third World, a kind of reparation for damage inflicted on the atmosphere.

The developed countries thus have the first responsibility to drastically reduce greenhouse gas emissions before they ask the developing world to slow down their emission rate or adopt new standards to reduce emissions.

Developing countries have taken a unified stand, that they are in no way responsible for the state of affairs, that their 'share' of the atmosphere has already been eaten into and destroyed by the affluent countries, and that if any measures are to be taken the rich countries must come up with the money and make available the technology to help them implement measures. Their responsibilities will be limited to the extent that these funds and technology are available.

ARRESTING GLOBAL WARMING

The globe is now 0.6°C warmer than it was during pre-industrial era and if this trend persists, by the turn of this century, it could even be 3°C warmer. Half of the carbon dioxide released by industrial processes is being permanently trapped in the atmosphere.

A part of solar radiation is trapped near the ground by the gases such as carbon dioxide, nitrous oxide, methane and other chlorofluorocarbons (CFCs) which have high heat-absorbing properties. These result in the greenhouse effect, (like the glass in a glass house preserving heat) causing the planet Earth to be warm and inhabitable. But excessive heat could result if these gases are in a large quantity and this could lead to a catastrophe. It is found that carbon dioxide contributes to 57 per cent of global warming, chlorofluorocarbons 25 per cent, methane 12 per cent and nitrous oxide 6 per cent. Future warming to which the earth will be subjected to will arise mainly from carbon dioxide and CFC emissions. Carbon dioxide is released mainly from coal, fossil fuels, oil, natural gas etc. which are the main sources of energy for industries. Chlorofluorocarbons is released from refrigerants, coolants, aerosol etc.

Global warming could cause severe upheavals in the world's weather systems. Frequent storms and cyclones ocean expansion due to heat and ice melting at the poles could cause the sea level to rise and this would flood many of the fertile deltas and that tracts of land. Rainfall patterns would alter, causing, parched soil conditions in some countries. Human health will be rendered more disease-prone due to increase in microbes, parasites and insects that thrive in warmer conditions. Since the natural eco-system can only adapt to a very small temperature rise, many plants and animal species will disappear.

The globe, now facing a common problem needs urgent effective action to counter it.

The global energy growth rate is between 2 per cent and 3 per cent annually; about 1.5 times the current rate of global population growth. The developing countries energy use in the year 2000 could be four times that of 1980, but, its legitimate need to increase and sustain economic growth to raise billions out of poverty outweigh the increasing use. Also, compared to the rich countries who are prodigious users of petroleum-based fuels, the developing countries' energy based activities are negligible. The North with 20 per cent of the earth's population accounts for 85 per cent of the global consumption of non-renewable energy.

The vegetation in oceans and forests absorb carbon dioxide. These are its natural sinks. The oceans may take up less man-made carbon dioxide from the atmosphere than has so far been thought. Hence there is a growing importance for implementation of afforestation programmes in both developed and developing countries.

Water Pollution is due mainly to the discharge of wastes into natural water. The discharge of huge amount of industrial wastes and domestic sewage are mainly responsible. The reduced fish catch in most of the rivers is only one obvious indicator of the extent of the pollution.

Industries such as the tanneries pollute the groundwater. The salinity of both the surface and undergroundwater has been raised to such levels that the water has been rendered harmful for irrigation and drinking. The water sources have been affected by discolouration and obnoxious odours apart from salinity and high chromium levels.

Air Pollution : Since air pollution is due mostly to exhaust industry and automobiles. It is more pronounced in urban and industrialised areas. As far as the rural areas are concerned it is mining that is primarily responsible for the air pollution since most mines are located in remote rural areas. What industry can do to an entire population is well exemplified by the case of Union Carbide in Bhopal. This single industrial accident left thousands dead and over 2 lakhs maimed for life. Due to heavy toxic emissions both from industry and from automobiles, cities have become a very unhealthy place to live in. Tuberculosis and other respiratory ailments are the city's biggest killers.

Forest and land resources are not the only things destroyed by overexploitation. It is also profit motive that has resulted in destruction of ocean resources and impoverishment of the fishing communities that are dependant on it. The repercussions of over-exploitation of fish for increased exports are borne ultimately by the traditional fishing community. The introduction of mechanised fishing has completely disrupted the centuries-old well pattern of life and economy of the traditional fisherfolk.

Urban Slums and Environment : The urban slum problem is caused by various factors, — the "push" and "pull" factors being the most significant. The "pull" factor is determined by the high concentration of industry in urban areas. Scarcity of food and other sources of survival either caused by natural calamities or over exploitation of the natural resources are some of the important "push" factors. The rural unemployment created by the "push" factor sends people to the urban areas, creating in the process a strain on the existing public amenities in the urban areas such as housing, sanitation, water and health.

The degradation of natural resources is also interrelated to the linkages that exist among land, water, air. For example forests are linked to soil conservation, water system and the over all atmosphere. Similarly the exploitation of these resources is linkd to the dominant social, economic and political forces and structures. It is only in the context of this global view with all its linkages and ramifications that one can understand better, the process of environmental degradation and its relationship to population.

BIODIVERSITY ISSUES

Key Issues

The key issues are: affirmation of sovereign rights of nations over resources located in their countries, linkage between access to these bio-resources and bio-technology that flows from research and development using these resources, and finally the issue of restructuring the financial mechanism through which new and additional funding.

Sovereign rights of States over their natural resources, the authority to determine access to genetic resources rests with the national governments and is subject to national legislation.

The other critical issue is of linkage between access to genetic resources and access to biotechnology flowing from it, a significant breakthrough for the developing countries. When developing countries provide genetic resources, they in turn be provided "access to and transfer of technology which makes use of those resources, on mutually agreed terms, including technology protected by patents and other intellectual property rights.

Access to and transfer of technology to developing countries be provided and/or facilitated under fair and most favourable terms, including concessional and preferential terms where mutually agreed, and where necessary in accordance with the financial mechanism established.

Patents and other intellectual property rights may have an influence on the implementation "and, therefore, as safeguard, countries must cooperate" to ensure that such rights are supportive of and do not run counter to the objectives.

BIODIVERSITY PRESERVATION

The term 'biodiversity' is relatively new. The concept of conserving biodiversity, encompasses genes, species, habitats and eco-systems, the importance of entire ecosystems and the living organisms therein has to be fully appreciated.

Today, of the estimated 10 million species wordwide, only 1.4 million have been named. Each year new species are being discovered, or rather, identified. At the same time, hundreds of species are also being destroyed, some without ever having been recorded. In the last 400 years, at least 700 known vertebrates, invertebrates and vascular plants have become extinct. In the U.S., along 34 species of plants and vertebrates have become extinct in the last decade even as environmental groups were demanding to take steps to protect them. Even if all-out efforts are made today worldwide to protect the planet's biodiversity, and to find the enormous resources required to do so, many more species of plants and animals would die out in the interim.

The reason is that 'progress' as mankind has known it so far cannot suddenly be halted. Thus, there is an on-going process of habitat loss and fragmentation, the air and water continue to be polluted, living resources are being over-exploited almost everywhere, in the rich and the poor world, forests are being cut down, both temperate and tropical, new genetically-manipulated species of crops are being introduced to enhance food production and the accumulation of carbon dioxide and other gasses is leading to climatic changes which could further devastate these living resources.

While some of this can be slowed down, it is evident that there has to be a clear understanding of how and why the process of destruction has begun and only then can strategies to reverse it be devised.

While the industrialised world is increasingly aware of the dangers posed to the world by the depletion of tropical forests, as crucial sinks for the carbon accumulating in the atmosphere through the burning of fossil fuels, much of this has taken place due to the high consumption of wood and wood products in that part of the world. For instance, the imports of tropical hardwoods in the Netherlands increased five-fold between the Sixties and Eighties. The Japanese have been systematically cutting down forests in South-East Asia to meet their demands for wood-based products, such as disposal chopsticks.

This period coincided with the increasing indebtedness of developing countries, a drop in commodity prices and the inevitable compulsion for them to sell more of their natural resources to remain where they were economically. The connection between the reverse net capital flow between the southern and northern countries and the depletion of biodiversity in the southern nations, which are the main repositories of these natural resources, cannot be ignored.

Furthermore, regardless of national policies or environmental consciousness in the poor countries, the power of transnational corporations to use the economic situation to encourage over-exploitation of natural resources is now a well-recorded fact. For example, in Papua New Guinea, foreign timber companies were roaming the countryside with the self-assurance of 'robber barons': creating social 'disharmony' and ignoring laws in order to gain access to rip out and export the last remnants of the province's valuable timber.

It is hardly surprising that third world countries are fairly paranoid about holding on to their sovereign rights over their natural resources, whatever, their value to the globe.

They now realise that the promises of higher food production through genetically improved seeds, sold to them by the West along with the whole packet of 'green revolution' fertilizers and pesticides has been a mixed blessing, at best. Not only has the introduction of high yielding varieties led to a proliferation of pests, and the destruction of large areas of standing crop but even hardy natural species that had survived over centuries have been destroyed. In Indonesia for instance, the introduction of hybrids in the last 15 years has led to the destruction of an estimated 1500 local rice varieties.

Given this state of affairs, countries in the North and South have been forced to acknowledge that there is an urgent need to preserve what remains of precious natural resources and to check further destruction. This requires both *in situ* conservation methods, that is, creating protected areas so that species can survive in their natural habitat, and *ex situ* facilities, where genes can be isolated and preserved until such time as their full value is realised. While the former can be done by poor countries, the latter requires technological know-how which has been developed in the North.

Until recently, without appreciating the value of their biodiversity, many poor countries have allowed Western scientists' access to their lands and forests. As a result, irreplaceable genetic resources have been taken, without any payment, experimented upon, and sold as new seed or new drug to these very countries at a price which includes the so-called incentive to the inventor or Intellectual Property Rights (IPR). Today developing countries are demanding not only payment for genetic resources, which they claim have been preserved and developed over centuries by their people, but access to the technology that turns them into a commercial commodity.

Biodiversity : the Need to Conserve

The growing pressure on land from population and urbanisation, from the needs of agriculture to feed the growing population, and from demand for fuelwood and fodder and timber have resulted in largescale deforestation and the shrinking of the natural habitat of thousands of species of plants and animals, threatening them with extinction.

Biodiversity means the immense variety in forms of life on this earth considered by scientists to be a precious natural resource which must not be lost. A species which becomes extinct is lost forever and obviously cannot be reproduced in a laboratory. Over the last hundred odd years more species of plants and animals have become extinct than ever before and that if some urgent steps are not taken mankind will lose this precious heritage.

The preservation and conservation of this biodiversity has become one of the key issues. To conserve this diversity the natural habitats where the maximum number of these species are to be found, the thick green tropical forests, must not be allowed to be destroyed any more.

The recent advances in genetic engineering which ushered in the green revolution was made possible by the such variety of wild species of foodgrain plants for the purpose of cross breeding. Genetic engineering uses genes with certain specific properties to improve the known varieties of food crops. If these wild varieties are lost and become extinct, then that genetic pool with its characteristic qualities will also be lost.

Plants growing wild in the forest may hold the key to cures for many kinds of diseases. A large number of medicines have a herbal base. The neem tree, could well replace pesticides used in agriculture. 'Neem' berries can be as effective against pests without being toxic to man and animal.

There is no difference of opnion between the developed and the developing countries on the need to preserve biodiversity. The poor countries are willing to help in preservation and conservation — the tropical forests are mainly in the developing countries — but since they do not have the funds to do this, they should be given global assistance. Access must be provided to the biotechnology that may be developed using these bio-resources.

Till now the developed countries often had free access to the bio-resources (say, the plants growing in wild) of the Third World, but technology developed by using these resources was patented and protected under intellectual Property Rights. If there was a major breakthrough in research, the country which had given access to the plants got nothing while the rich countries which invested in research made huge profits and resold the technology to the Third World.

Biotechnology flowing from use of these bio-resources must be made available to the country which gives access to the basic genetic materials. And, profits from this biotechnology must be shared between the country where the plants originate and the country which develops the biotechnology, whether in the area of medicine or new varieties of foodcrops.

Geographical areas which are considered to be the natural habitats of plants and animals threatened with extinction may be identified and nations within whose jurisdiction they fall be asked to take measures to conserve these forests.

Developing Nations are rich in their biological diversity the world needs to conserve. We should take immediate steps, e.g., cutting subsidies for energy, irrigation or logging, that would profit us even before counting the green gain. In reconciling growth with greening, all countries have responsibilities as well as rights. But it is a disappointment that a path to a binding forest treaty was not achieved. The countries with tropical forests should develop them in a sustainable way. The Moreton Bay chestnut tree in New South Wales and Queensland in Australia, the black mulberry, and the Madagascar periwinkle, may all have ingredients in them to inhibit the AIDS virus, the scourage of today threatening whole swarms of population. The old English yew tree, tree of life to the pagans may have in its many compounds some which could be developed into tumour-fighting medicines; while it has been known for some years that a substance called 'taxol' in the bark of the Pacific yew has properties for fighting against breast and ovarian cancer. Plants, some of which have appeared worthless to human needs, can suddenly be found to harbour compounds for combating serious diseases. All this merely emphasises the indispensability of wild plants for human survival.

'Biodiversity' is the variety in animal and plant life, which form the natural world. This world is our living laboratory, as well as our larder, medicine chest, raw materials store, and playground. By impoverishing it we seriously impoverish ourselves and our descendants.

Every 100 hours, there are one million more births than deaths in the world. Whether or not the world achieves a biological sustainable economy will be severely determined.

ACCESS TO GENETIC MATERIAL

Access to genetic material to researchers should not deny the fruits of biotechnology to the country where the genetic material originated.

Issues of bio-diversity must be linked with the sharing of biotechnology, sharing of results of R and D and the commercial benefits derived from bio-materials.

Often it is traditional knowledge acquired by communities over centuries that forms the starting point for identifying particular species for particular applications and that "this must be kept in mind regarding intellectual property rights related to biotechnology.

In India, over 50,000 varieties of rice are found and this offers a tremendous base for future research. While India recognises that biotechnology research must be adequately compensated, mechanisms other than the prevalent system of IPR are required. The researcher can be compensated but without restricting the use of the biotechnology.

When the original genetic material is collected, no payment is made to the country. However, the technology derived is patented and often sold back to the very countries from where the original material is taken. The countries of the South are rich in biodiversity and it is this material which will form the basic genetic pool for future research in the area of food and medicine.

So far the countries of the North have been pressing that areas rich in bio-diversity are preserved (for example the tropical forests), that these are declared to be global resources with free access for all to the bio-material. And naturally, the developed countries would like to patent any technology that is derived from the use of this material.

If free access to bio-material is given by any country, then the technology derived from it must also be made available to it free of any patent right, and free of cost.

Any safe global environment strategy must get the co-operation of the developed as well as the developing countries and issues that need to be resolved and steps that have to be taken must be seen in a spirit of a global partnership.

For this, acces to environment friendly new technologies was essential so that more countries are not forced to buy older and more polluting technologies that are causing havoc with the planet's environment. Mechanisms must be in place to prevent wrong decisions for lack of funds, and global funds must be found to subsidise the difference.

A global fund would take care of all incremental costs incurred in switch over to environmentally more friendly technologies.

SAVE THE EARTH MISSION

Cure the Earth in Danger

The earth has become a very sick planet and urgently needs a cure. A disaster is looming unless everybody joins hands to take on the environmental dangers and to find ways to halt the march to mankind's destruction.

The environmental pollution which the people the world over have inflicted on this planet has been contributing increasingly to the destruction of life on the earth. We should seek to explore ways of offering a solution to the growing dangers facing mankind such as the draining of resources, progress of environmental pollution and the explosive increase of pollution in the developing countries.

There are grave environmental problems brought about by a combination of factors, among which is industrial pollution in highly industrialised and mass-producing countries. There are shortcomings of one sort or another in both industrialised and developing countries and in recognising, facing and adequately planning to meet the growing environmental pollution problems. This is the gravest warning of a coming disaster unless all people of the world join hands to squarely meet the environmental dangers and find quick and suitable ways of overcoming what would otherwise be a march down the path of destruction of mankind. The earth had become a very sick planet and needed urgent cure.

Continued economic growth, together with widespread mismanagement of resources, is having an explosive impact on the environment everywhere. Pollution explosion and the use of energy posed other dangers. If the right effort was made to stop these dangers, the globe could be saved in, say, 20 years; otherwise, there would be a global collapse.

The world's population was 3.2 billions in 1972 but, in just 20 years, it had exploded to 5.6 billions and is sure to reach over six billions by the end of this century. Pollution continues to increase, endangering the lives of people the world over and adversely affecting food production, particularly in the developing countries. During the last decade no less than 60 countries have had a negative growth in production.

Human use of many essential resources has already surpassed sustainable limits. The environmental pollution is also contributing in various ways to the growth of poverty and this can be avoided if a check is put in such areas as uncontrolled production, population growth etc.

The world is facing more than one such crisis. Ozone depletion is one of them. The global ozone levels may decline by a dangerous 20 per cent by the year 2000 A.D. Contributing to this have been such factors as industrialisation, tourism, resources mismanagement and urban development. Man is sick in body and spirit, given to egoism and sheer consumerism, and these have contributed to the crisis of values.

Drylands the world over are becoming drier due to the extensive land use of technological, industrial and other purposes. The dryland problem is a problem of the poor and some 900 million people have been affected world-wide. We should recognise how the short-sightedness of the modern man has contributed critically to the destruction of the environment. The role and responsibility of the young all over the world in adapting human behaviour to the requirements of a harmonious life with the rhythms of nature are important. The value systems on which the relentless pursuit of materialism has had high priority need to be reconsidered.

On such issues, the tradition and the achievements of Japan might provide a model for the World community by setting a model of local action through continuous efforts of citizens, local authorities and corporations working in concert to address local environmental issues.

The apparent conflict between development and environmental conservation demonstrates the mutual vulnerability of both the developed and the developing countries. Future harmony in the world necesitates joint action and partnership.

We should realise true symbiosis with nature, which will be in harmony with ecology. "We should be prepared to accept the cost which any action friendly to earth would entail. Otherwise fatal pain is sure to befall the whole of mankind".

SAVE AND ENRICH THE EARTH AND ENVIRONMENT

The time has come for all of us, the Government and public to get together, to let nature restore itself by preventing pollution and environment decay.

Today the scene is frightening. Forests are being denuded, lakes are filled up for development, villages are devastated, cities are over-built and air we breathe is polluted with increase in carbon dioxide, sulphur dioxide, etc. Factories, Mining activities, Vehicles we travel in and the houses/layouts we live in are all contributing to this disaster in the making. We should stop denudation of forests, help conservation of water, ensure proper pollution controls in Industries, Manufacturing activities, Sewerage handling and disposal units etc., check vehicular emissions of injurious gases.

Today there is not a single area of this planet, no matter how remote, that is untouched by pollution, key issues are desertification, marine and freshwater pollution, the destruction of habitats and loss of wildlife — underdevelopment and cripping poverty are the greatest cause of environmental destruction.

The concept of development that is environmentally sustainable is the need of hour. The gulf between what we knew about the state of the Earth and what we were actually doing about it had grown steadily. With this laissez-faire attitude disaster lay ahead. The cumulative effects is crushing.

Environmental issues are high in people's minds and lives. But, recession pushed bread-and-butter issues forward and relegated the environment to the background.

The primary goal should be to lay the foundation for a global partnership between development and more industrialised countries, based on mutual need and common interests, to ensure the future of the planet. We need to find a viable and equitable balance between environment and development. A new realism must be created and maintained that the industrial nations commit themselves to environmental aid transfers to the development by huge sums of money.

Let us look at underlying patterns of development which cause stress to the environment, poverty in developing countries, levels of economic growth, sustainable patterns of consumption, demographic pressures, and the impact of the international economy. An Earth Charter should embody the principles to govern the economic and environmental behaviour for action in major areas affecting the relationship between the environment and the economy; the means to carry out this programme by making available to developing countries the additional financial resources and environmentally sound technologies they require to fully participate in global environmental conservation; integrate environmental considerations into development policies; strengthen institutions to implement these measures; and conventions on climatic change, biological diversity, and forests. We should endorse the principle of 'green growth' that our future prospects depend on sound environmental management now.

Protecting the environment is not a favour the poor nations do to the rich ones in return for money. The poor nations already suffer from degraded environment and have fewer defences against further degradation. But the developing world has a strong case in seeking Western cash and technology to avoid the worst environment traps of economic growth, and the West has an obvious interest in seeing that this does not happen. The growing size of the global population is alarming. Whether the number will be twice or three times the present 5.5 billion will depend largely on how the unmet demands for family planning round the world are met. Family planning contributes immeasurably to the solution of environmental problems. It is not disputed that most people all over the world seek to have small families.

MAKE EARTH HABITABLE

The Earth is a fine place to live. The air is breathable, temperature and pressure are just right, and the atmosphere screens us from ultra-violet radiation from the Sun which would otherwise fry us in minutes. Did this match between Earth's environment and life, and did this come about because life on Earth adapted to the condition it found? Or did life alter the Earth's environment to make it — and to keep it — habitable?

We know that many of the conditions we and other living beings need to survive are produced by living things themselves. Nearly all the nitrogen and carbon dioxide in the atmosphere today have biological sources, though in the early history of Earth they were produced by volcanoes. So living things must sometimes have altered the environment as well as adapted to it. The greenhouse effect of gases in the atmosphere keeps the Earth warm by trapping some energy when sunlight which has been absorbed by the Earth is re-radiated back into space Since the mixture of gases on Earth would be different without life, so would the temperature of the Earth. Our planet could easily be either too hot (like Venus) or too cold (like Mars) for life to exist at all. Nonetheless, despite the fact that the Sun has long being getting hotter and the Earth has been many huge meteorites, life has thrived an unimaginable length of time — 3.5 billion years.

The "Gaia" hypothesis seeks to explain the survival of life on Earth of billions of years by treating life and the global environment as two parts of a single system. The system sometimes also called simple "Gaia" has developed so that it can regulate and repair itself. Regulation means life actively keeps the global environment comfortable for life to continue. And, says the hypothesis, if Gaia is knocked dangerously off balance, it can repair itself.

Consequently, the system as a whole is fairly immortal, able to survive even major crises such as the impact of huge meteorites — something which has probably happened many times during the history of life on Earth. This does not mean that any individual species, such as human beings, is guaranteed to survive. It does mean that life as a whole is very robust — particularly simple bacteria which until one billion years ago were the only type of life on the planet.

Regulation and Repair

The Earth on which life first appeared was totally unlike the planet we know today. The planet was only just past the final stages of formation from a nebula of gas, dust and rocks. Cataclysmic impacts with planetesimals, huge meteorites tens of kilometres in diameter, were still occurring frequently. Probably, the composition of the atmosphere was dominated by carbon gases thick carbon dioxide, methane. It certainly contained no oxygen.

The Sun was also different. Assuming that it has developed in the same way as other stars of its size and composition, it must have been 24-30 per cent less hot than it is now. If one were to turn down the Sun this much today, the Earth would cool so much that the oceans would freeze. But we know this has not happened since life appeared because practically all life requires temperature above freezing to thrive. Warm-blooded animals which regulate their own temperature only evolved comparatively recently.

The common explanation for this paradox is that the greenhouse effect of the early Earth was stronger, probably because of the greater amount of carbon dioxide. Overtime, the greenhouse have been reduced in the atmosphere, and the Sun has become brighter, so that the temperature at the surface has remained comfortable, for life. Gaia's supporters believe this is evidence for the existence of ''Gaian'' regulation mechanisms — that life on the planet actively regulated the temperature to make this happen.

The second idea is that of 'repair'. If Earth was a peaceful place there would be no need for 'Gaia'. But Earth is not peaceful. It is continually bombarded by fragments of rock from space. Almost all the rocks are small and burn up in the atmosphere to form shooting stars but a few are big enough to reach the ground. About once every 100 million years nowadays a planetesimal may hit the Earth, shooting enough dust and gas 15 kilometres or more into the atmosphere to block out the Sun over most of the planet. These cosmic collisions are thought to do enormous damage, wiping out huge numbers of species.

But life survived these catastrophes. In fact on a geological time scale, new species arose quickly to replace the old one. Supporters of

'Gaia' argue that this is evidence in support of the hypothesis. They say it shows that joint system of life and environment on Earth is both robust and able to repair itself quickly. Though biological control of the global environment may breakdown immediately after such disaster, life rapidly regain control after these events and begins to regulate again. This does not mean that planet is left unchanged. The species on Earth would be different after the catastrophe, causing the new environment to differ too.

The Carbon Dioxide Pump

One of the results of the 'Gaia' hypothesis is that scientists have looked more carefully at the mechanisms which regulate the Earth's environment. It is early to theorise about these mechanisms but two such theories demonstrate well the complexity of the Earth's processes and the difficulty of studying them. Neither of these theories depends explicitly on the 'Gaia' hypothesis, but both were to some extent inspired by "Gaian" thinking.

We believe the amount of carbon dioxide in the atmosphere has controlled the temperature of the Earth throughout its history. If 'Gaia' exists then we would expect that atmosphere carbon dioxide was under biological control.

Over millions of years, the amount of carbon dioxide in the air is controlled by a balance between the rate at which it is released by volcanoes and the rate at which it is taken up by the chemical weathering of rocks containing silica. Neither weathering nor the eruption of vulcanoes apparently involves life, so how can life control atmospheric carbon dioxide?

The weathering of silicate minerals, found in granite for example, takes place in the soil and uses up carbon dioxide. Plants and microorganisms such as bacteria absorb carbon dioxide from the air and pump it into the soil either through their roots of by causing decay in the soil. So typically there is about forty times as much carbon dioxide available in soil as there is in air, and this speeds up the weathering. As a result, more carbon dioxide ends up in the soil and less in the atmosphere. Eventually, carbon is laid down as carbonate rocks such as limestone.

An increase in the Earth's temperature would probably expand the tropics at the expense of polar regions. This would increase the amount of biological activity on the earth and the amount of carbon dioxide pumped into soil. Consequently, carbon dioxide would be removed from the atmosphere, reducing the greenhouse effect and tending to reduce the original rise in temperature.

This is a "Gaian" temperature regulation mechanism. However, some scientists have proposed purely non-biological mechanisms for temperature regulation by way of atmospheric carbon dioxide. For example the weathering of silicate minerals by carbon dioxide will go faster at higher temperatures, and the increased rainfall associated with this increase in temperature would also tend to accelerate the inorganic reactions. Proponents of 'Gaia' argue that organisms process and cycle carbon so rapidly that such inorganic feedbacks would be ineffective compared with biological mechanisms.

In a cloud, every water droplet starts life by condensing around a microscopic particle of water absorbing material called a cloud condensation nucleus of CCN. If there is shortage of CCN, as sometimes occurs in very clear air, then, clouds form with fewer but, larger, water droplets. If there are plenty of CCN there will be more but smaller cloud droplets. Smaller drops make of whiter clouds and reflect more of the Sun's light back into space, thus cooling the Earth. Potentially, therefore, the climate of the whole planet can be altered by changing the number of CCN in the atmosphere.

It is possible that over large parts of the oceans the main source of CCN is dimethylsulphide gas (DMS) emitted by algal plankton that live close to the surface. Sunlight changes the DMS into acids which form CCN. Scientists have calculated that the algae release more sulphur in DMS than all the power stations in the world. And the pattern of CCN over large areas of the oceans is roughly consistent with their being derived from DMS.

According to the "cloud-algae" theory, the algae exert an important influence on the temperature of the Earth through their effect on the whiteness of clouds. The emission of DMS thus helps to keep the planet cool — another 'Gaian' like regulator.

But even if the algae do influence global temperatures in the way, it is not clear how strong the influence is or whether it is a regulatory mechanism. We do not know whether algae produce more or less DMS as temperatures rise. For the whole system to tend to stabilise the temperature, the cloud algae feedback has to be negative — DMS production and the whiteness of clouds must rise as global temperatures go up.

Recent measurements made in Antarctic ice up to 15,000 years old suggest the opposite. As the temperature increased at the end of the last ice Age, the amount of DMS derived sulphur in atmospheric particles seems to have fallen. This would mean that the relationship between clouds algae tends to destabilise global temperature. This seems inconsistent with Gaia. However, it is possible that the sudden warming at the end of the ice Age represents a temporary break down of the normal regulators.

LANDSLIDES MANAGEMENT

Landslides are a primary cause of damage to many hillside structures including roads, buildings, bridges, cultivated lands, forests and orchards and degradation of hillside ecology. Understanding landslide mechanisms and collection of data can help choice of landslide management technologies and enhance the ability to control landslide hazards to achieve an improvement in the quality of life in the hill areas.

The use of prestressed anchors to improve the stability of rock slope and horizontal drains for stabilising deep seated landslides is recommended.

The incidence of landslides in hill ranges is a common phenomena. The variation in the degree of incidence of landslides in the various hill ranges can be from very high to low.

A concerted effort to mitigate landslide losses through landuse management in hill areas, adoption of appropriate landslide control techniques, monitoring landslides using subsurface and surface instrumentation, early warning systems, hazard zonation maps and dissemination of correction techniques can significantly reduce the damage as well as contain landslide hazards.

Rock Slope Stability

We must understand the mechanism and causes of landslide in rock slopes and the constituent materials of the hill slope areas from geological considerations and the existing nature of the hill slope and the surrounding environment for developing appropriate measures to control different types of landslides. Different techniques to control the landslides and the remedial measures must take into account the mechanism of the slide.

Rockfall

The rockfall prone hillslope areas comprise predominantly of grey highly jointed medium to thinly bedded slates. The hill slopes are very steep with slope angles ranging from 50 to 70 degrees or more with uneven slope surface. On the higher reaches, the frequency of the joint planes increase resulting in a decrease in the size of the individual rock fragments. The rockbeds in the formations are intersected by numerous joint planes. Small gullies developed on the slope surface act as passages for the flow of the debris down the slope.

The hillslope material deteriorates and accumulates within the oval basin at the top and falling down on to the road from a slope height of more than 100 m with a high velocity. The rockfall prone area is slowly extending in the uphill direction and would endanger the existing high voltage transmission tower on the hill top in due course of time in the absence of effective remedial measures.

The prime source of rockfall is (i) shattered rockbeds along the periphery (ii) accumulated loose debris in the oval basin at the top and (iii) other fragments randomly scattered over the slope area. We should mitigate damage from rockfall hazards by extensive use of geogrid and protecting the affected slope areas and the accumulated loose debris at different places.

Landslide of Sorts

The landslide affected areas comprise mainly of biotite gneiss rock. The thickness of interbanded biotite layer range from a few mm to two cm and the rock intersected by several sets of prominent joint planes. Two types of slope failures viz., planner type and rockfall/topping type are observed.

To predict the mechanism of the slope failure, several sets of joint planes are studied with the help of plotting on the stereonet. Samples of guages materials within the joint planes and the soil in the slope areas are tested for stress/strain behaviour. The inherent lack of strength is respcnsible for the development of tension cracks on the hill scope. These cracks allow rainwater into the hill slope and enhance the sliding activity.

The remedial measures for rockfall/topping in this area include back cutting of the rockmass, removal of the hanging rock blocks, rock bolting technique and sealing of the tension cracks.

The hillside formation at landslide consists of states in a highly shattered condition following interaction of the various sets of joints. The rockfall takes place due to steep nature of the hillslope having inclination of 45° or more. Due to further denudation of the hill slope area, the probability of rockfall of size 20-30 cm increases.

The prime source of rockfall is identified in a few cases. A slope of about 150 sq.m. area is covered with a mantle of synthetic geogrids. Trench bolting and surfacial bolting are adopted. The aim of putting the geogrid is to stop the rockfall activity by keeping the loose rock fragment intact and to promote the possibility of growth of local vegetation on the rocky surface which in turn, will further increase the stability of the hill slope in the near future and reduce the rockfall activity in this area.

The hill slope is composed of dark grey to black slate of variable thickness ranging from a few cm to 50 cm. The downhill slope experience significant amount of subsidence. The mass movement is primarily being incited due to culminated effect of toe erosion and the interaction of the planes of discontinuities.

To improve stability of the hill slope, restraining structures are recommended in addition to the measures adopted for the surface erosion of the slope.

Computer Programme for Rockfall Trajectory

Rockfalls are a form of slope movement and constitute a serious hazard to the public and abutting structures along the mountain highway. If the trajectory of the rockfall can be predicted, it is possible either to shift the highway to some distance or to construct a barrier to prevent the falling rocks from causing damage to nearby structures.

A computer programme is developed to predict the possible rockfall trajectory. The objective of developing the programme is to generate a set of possible rockfall trajectories for various conditions of slope geometry and frictional properties. Such data would enable the location of structures in a zone free from rockfall hazard and to install rockfall control systems at proper locations.

The programme may be extended to accommodate frictional, rolling of sliding of the bounder along the slope and to include graphics to get the pictorial view of the rockfall trajectories.

There is need for instrumented monitoring of landslides, collection of relevant field data and developing suitable models to forecast the behaviour of landslide area as well as to develop effective and economical corrective techniques.

The present approach to deal with slope stability problems does not adequately take into account the various factors whose interaction cause instability of slopes. Majority of the landslide incidences fall in the category of rainfall induced landslides, especially in the areas subject to a limited period of intense monsoon but dry during rest of the year.

In such situations it is necessary to study a number of factors such as rainfall, run-off (vs.) percolation ratio, seasonal change in moisture content, degree of saturation with depth, influence of these parameters on the strength in saturated as well as unsaturated conditions, related factors and incorporate the above factors in the overall evaluation of the slope stability to forecast the behaviour of landslide area.

Deep Drainage Techniques

Stability of saturated slopes can be improved significantly by lowering the piezometric level. This can be realised by installing horizontal drains or deep trench drains.

Horizontal drains consist of 4.0 to 60 mm acorss PVC pipes with slots or holes driven at a negative gradient of 5-10 degrees into the hill slope. The pipes drain water from deep inside the slide mass which can be collected and discharged. Deep trench drains consist of deep trenches excavated into the slope and back filled with clean and well draining rubbles. The zone of high permeability so provided helps to drain water from the slide mass. It would be advantageous to provide a geotextile layer covering the rubble drain. The project aims at installing horizontal drains and/or trench drains in a slide area and studying the efficacy and economy of these deep drainage techniques in stabilising landslides.

The incidence of rockfall generally occurs after rains. Depending upon the type of failure in the particular stretch, the following remedial measures are suggested :

— Use of coir netting to promote vegetation on the denuded slopes areas.

— Removal of rock/boulder physically.

— Providing Nelton mesh for rockfall control.

The factors emerging as primary causes for a recent case of incidence of landslides are :

The road being aligned through the steeper terrain which is largely composed of colluvium.

Strata encountered at many locations being in a relatively weathered state and the hillside morphology is such that a significant portion of the rainwater would run down along the surface on which the road is aligned.

The combination of above factors results in a debris slump to debris flow type of landslides and rockfall. In some locations thin colluvium mantle is overlying the rockbeds. This mantle material is partially or fully eroded by the rainwater leading to the rockfall incidences.

The remedial measures required at the landslide areas include provision of restraining structures at a few location, encouraging the growth of vegetation on the denuded hillslopes and installation of rockfall barriers.

FALLOUT FROM EARTH SUMMIT

Further Road to Tread from Earth Summit : This should obviously take us on to forestry principles on environmental policy and development initiatives and programme of action into the next century in areas like poverty, pollutation and pollution.

The world has radically to change its approach to economic growth and the problems of underdevelopment.

Forest principles should keep local community interests firmly tied in with programmes to protect forests.

The richest 20 per cent of the world enjoy 82.7 per cent of the income and the poorest 20 per cent of the have only 1.4 per cent of the global income.

The issues are: consumption patterns, gross disproportion of consumption to population levels and these are vital issues.

The real culprit of environmental degradation is "the character of development which has concentrated of heaping prosperity on a small proportion of the population by indiscreet and inequitable use and command of resources for its lifestyle". The poor have no part in shaping development policies which continue to be centralised and top down in its approach. It is the 'vibrant power' of grass-roots institutions they have to be allowed to flourish, for ecology and development have to be defended on a 'scale and depth and by the masses of people themselves'. 'Towards green' villages does not mean only environmentally sound rural development as one of only trees and watershed but "more of village institutions, laws, financial frameworks and planning processes, the key elements of not only state policy but also a national framework for people's action for environmental management and regeneration in rural areas". It is also essential that our planners keep in focus the most economically disadvantaged and ecologically devastated segment of the population such as rural women for example, whose strenuous lives are already stretched beyond their fragile 'carrying capacities'.

Ultimately it all comes down to the reality of micro planning prioritised above all other considerations and vested interests so that we avert the tragedy of economic and ecological devastation that constantly looms over this fragile, overburdened planet.

THE EARTH SUMMIT — STEPS FOR SUSTAINABLE DEVELOPMENT

The principal environmental and development challenges and opportunities and their interlinkages are : Poverty in developing countries, unsustainable consumption patterns, demographic pressures in a period of rapidly rising world population and the impact of international economy. Action is required on :

— Protection of the atmosphere.

— Conservation of biological diversity and environmentally sound management of biotechnology.

— Protection of freshwater resources.

— Protection of oceans, seas and coastal areas, and the rational use and development of their living resources.

— Environmentally sound management of hazardous wastes (including toxic chemicals).

— Prevention of illegal traffic in toxic products and wastes.

— Improvement in the quality of life and human health.

— Improvement in living and working conditions of the poor by eradicating poverty and halting environmental degradation.

An Earth Charter comprises of:

— Action Plan for the 21st Century.

— Conventions on climate change and biodiversity.
— Agreements on forests, institutions, financing and technology.
— New partnership between developed and developing countries.

Human beings are at the centre of concerns for sustainable development and the goal should establish a new and equitable global pertnership through the creation of new levels of cooperation among States, key sectors of societies and people.

The right to development must be fulfilled so as to equally meet the developmental and environmental needs of the present and future generations. An order to achieve sustainable development and environmental protection shall constitute an integral part of the development process and cannot be considered in isolation from it.

All people shall cooperate in the essential task of eradicating poverty as an indispensable requirement for sustainable development. The need to decrease disparities in standards of living among people and in that context the special needs of developing countries and those most environmentally vulnerable shall be given priority.

States should cooperate to promote a supportive and open international economic system that would lead to economic growth and sustainable development in all countries. Trade policy measures for environmental purposes should not constitute a means of arbitrary or unjustifiable discrimination or be a disguised restriction on international trade. Unilateral actions to deal with environmental challenges outside the jurisdiction of the importing country should be avoided.

One of the key issues is the unsustainable patterns of production and consumption of some States and their elimination and the promotion of appropriate demographic policies. Cooperation should strengthen endogenous capacity building for sustainable development by improving scientific understanding through exchange and transfer of technologies, including new and innovative ones.

The role of women, youth and indigenous people in environmental management and development is important. Another crucial principle is that at national level, each individual shall have appropriate access to information concerning the environment, including hazardous materials and activities in their communities.

The facilitation of public awareness of environmental issues by the State to judicial and administrative proceedings, including redress and remedy should be provided.

Compensation for Victims : The development of national law regarding viability and compensation for the victims of pollution and other environmental damage, cooperation among states to discourage the relocation or transfer to other states of activities and substances that harm human health and cause environmental degradation are also sought.

The principles should embrace the entire gamut of concerns that the world has to urgently contend with to protect the planet, and to set the stage for international agreements which would respect the interests of all and protect the integrity of the global environmental and development system.

FOLLOW UP OF EARTH SUMMIT (AGENDA-21)

Scientific predictions are that the hole in the ozone layer and consequent danger from over exposure to ultraviolet radiation from the Sun, will develop first near the two Poles and will have lesser effect on the areas between the two tropics. Harmful chlorofluorocarbons that destroy ozone should be phased out.

Some scientists have predicted that global warming also will affect the countries in the temperate ozones, closer to the Poles, more than the tropical countries.

Funds for environmental generation are to be treated like taxes on over consumption.

Japan today has a far more fuel efficient economy than the U.S. and it is for this reason that it has been pressing for stricter environmental regulations. It stands to gain billions of dollars of business from stringent pollution control rules.

Environment issues cannot be solved by delinking them from developmental issues. There is no way the poor can be taxed further for the environmental sins of the rich. They simply do not have the funds or the technology.

In the poor countries life has always been cheap. The affluent have everything to lose from a global catastrophe. It is they who are paranoid about the grave consequences of ozone depletion and global warming. It may be cynical to say that the poor are used to natural disasters, that floods, famine, hunger and thirst are part of everyday survival issues in this part of the world. Let them, for a change learn to cope with hazards of over-exposure to ultraviolet radiation and uncertainties of global warming. And if nothing much is done to take effective steps to ward off the predicted disasters, the affluent nations will have only themselves to blame. The poor will have only their debts to lose.

Long-term environmental productivity (biological sustainability) is generally in the best interest of the subsistence needs of the rural poor. However, it is not clear why rural people should naturally opt for long-term biologically sustainable strategies over alternatives that provide economic mobility. In other words, the goal of environmental protection is likely, in the near future, to position our environmentalists on the opposite side of the political fence from small rural producers. Development then becomes merely the act of substituting one environmental system for another. In such substitution, there must be losses and gains: residents of the site may lose their livelihood or gain a new start; outsiders may win access to a resource, or similarly lose access. However, those who partake in the construction always win. Whether in the developed or developing world, this latter point is axiomatic.

The environmental movement, and similar institutions that empower the voiceless concerns of society, are valuable and necessary components to social democracy.

Indian environmental activism is synonymous with rural social activism, empowering rural poor and providing them with capabilities to maintain control over their productive means. It is not a question of who wins, or who loses; who is right or who is wrong; but how far society moves in democratising the entire process of rural development.

In the case of large rural construction projects, technocratic power must be confronted before a proper system can evolve that guarantees fair project evaluation, proper representation of all parties and each long-term concern, and for a democratic route to dialogue and compromise and consensus which is so much spoken of these days.

Global warming (greenhouse effect and ozone depletion) are the two issues related to the very ecological survival of this planet, while the question of sustainable of the earth's population. The ecological survival and the economic development rightly belong to a single integral issue, a question of environment friendly development.

Sustainable development is that which meets the needs of the present without compromising the ability of future generations to meet their own needs. This means a cautious planning for future resources and a correction of the past misdeeds; that is, clean up the mess and "reduce and eliminate unsustainable patterns of production".

One should recognise the integral and interdependent nature of the earth. The condition that the right to exploit one's own resources should not affect the environment of another nation negatively; the principles include clauses for mobilising women and youth and tapping the traditional wisdom of the indigenous peoples for achieving the sustainable development for a better future.

Indigenous capacity-building for sustainable development by improving scientific understanding through exchanges of scientific and technological knowledge and environmental disputes peacefully and by appropriate means are parts of the principles.

The inherent destructive nature of warfare going against the sustainable development and the interdependence and the indivisibility of peace, development and environmental protection should be recognized.

Promotion of "appropriate demographic policies" should be pursued seriously.

The action plan for sustainable development for the decade and beyond to save the planet, the environmental measures aimed at a bright 21st century, are the major issues. The components, besides inter-linkages and means, include issues of population control, poverty combat, healthy consumption pattern, shelter, health, waste minimisation and recycling water management, sanitation pollution control, resource efficiency in land, energy, water, agriculture and forest, reforestation, anti-desertification programme, drought preparedness and relief, conservation of agencies, sustainable energy development, island development, marine resource management, managing radio-active and hazardous water enduring the participation of Non-governmental organisations, women, youth and others in environment protection, etc.

The major hitch is fund and technology transfer. The new packages should be entirely additional and delinked from any reviews or conditionalities on the national priorities and programmes.

Even the priority areas eligible for funding are also decided by the donors. These areas are: oceans, biodiversity, climate change and ozone layer. The amount of one per cent of the combined GNF of the industrialised North (OECD countries), is a must in addition to the existing average level of annual Official Development Assistance (ODA) of US $55 billion.

Not only that a commitment to US $ 125 billion is far from reality. The general call for new and additional resources to help the South in implementing the sustainable development programme would reduce the obligation for direct aid and instead force the South to go in for bilateral trade agreements with the developed nations which means deals on commercial basis and a circumvention of the polluter-pays-principle. In such an eventuality, when the global solution is tried at the local expenditure (of the South), the genuine concerns cannot be overlooked. It should be remembered that US $ 125 billion is the share of the North while US $ 500 billion is that of the South, because an annual expense of US $ 625 billion is estimated to be the cost of the environmental rebuilding and development in the South.

Forest eco-systems and resources are part of the national patrimony to be managed, conserved and developed by each country in accordance with the national plans and priorities in the exercise of its sovereign rights. Patents and Intellectual Property Rights (IPRs), technology transfer the environmentalisation of development are all important.

The genetic diversity itself is going to be threatened by the aggressive commercial genetic engineering. This is a possible threat to bio-diversity.

The modern pharmaceutical and drug industry depend on the tropical flora and fauna, as around 50 per cent of the drugs are produced from these natural bio-resources; globalisation of forests and minerals is a sophisticated attempt to further appropriate and colonise the wealth of the poor nations. The Convention on Biological Diversity was expected to include legally binding commitments to ensure the link between the access to the genetic material of developing countries and the transfer of biotechnology and research capabilities from countries as well as sharing of commercial profits and products derived from the genetic material.

Efforts to protect world's animal, plant and microbial species, forest development and an increase in international development aid would help.

The climate changes are partly caused by the human intervention. Forests, ocean, industrial emissions, population, agriculture and livestock are the major areas. Global warming and ozone depletion are two manifestations.

The chlorofluorocarbons (CFCs) cause holes in the stratospheric ozone layer, the protective cover of this planet that controls the ultraviolet radiations of the sun from entering into the atmosphere. Ultraviolet rays are understood to cause skin cancer, reduce the immune ability and affect the crop yields in the negative. In that case, ozone depletion affects both human life and food; it destroys biosphere.

The ozone depletion in the Arctic region has created panic among the Northern peoples. Mothers are not allowing children to play unprotected in the sun. The EEC, the USA and Canada are trying to face out CFC production at the earliest. Most probably by 1995 they shifted to new refrigerants. Already experiments are successful in blasting sound energy into cool waves.

Global warming or the greenhouse effect, the phenomenon that increased the temperature of the earth's atmosphere is caused basically by the oxide gas emissions of carbon sulphur and nitrogen. Carbon dioxide emissions due to fossil fuel burning is taken as the major cause and hence the interest in afforestation as it sinks the carbon emissions. The USA alone accounts for 24 per cent of the carbon dioxide emission of the globe.

As oceans form the major part of earth's surface and as water has a high heat retention ability than the dry air, ocean pollution can contribution much to the global warming.

The increase in Earth's surface temperature can lead to rise in water levels threatening the very existence of a few island nations, the livestock and rice cultivation. Fertilizer use and forest destruction, methane emission also are the reasons for the greenhouse effect. Bringing carbon dioxide emission back to 1990 level within a specific period is required. Instead of cutting down the emissions, overconsumption etc. be avoided.

BLUEPRINT TO SAVE THE EARTH

The comprehensive document has set the planet on a new course of sustainable development to save the fragile earth from environmental degradation. It is intended to provide a blueprint of action in all areas relating to the sustainable development of the planet from now until 21st century, 178 nations participated in the Earth Summit.

An estimated $600 billions is the incremental additional cost to make the overdue development of South sustainable. As much as $ 125 billions is needed annually in addition to this for economic assistance and the transfer of technology from the North. Debt relief measures should be kept under review so as to address the continuing difficulties of concerned countries.

Developed countries reach the accepted target of 0.7 per cent and augment their aid programmes in order to reach that target as soon as possible. Sustainable Development in all its complex multidimensional aspects covers population policies, health care, education, women, youth and indigenous people including local communities, and their rights to achieve livelihoods.

Food, security, clean water and sanitation and related issues of consumption patterns, wastage, demographic dynamics, human health and promotion of sustainable human settlements are also vital. The protection of freshwater sources, oceans, the safe use of toxic chemicals, the managing of hazardous and radioactive wastes sustaining biological diversity are also involved.

Other subjects include atmosphere, halting the spread of deserts, the conservation and rational use of forests and integrated approach to land resource use, issues such as environment-friendly technology, building national capacity of sustainable development, creating environmental awareness and the promotion of international legal instruments and mechanisms are also important.

BRIDGING THE GAP

The key to this issue is embedded in the concept of environmentally sustainable development. No one in the industrialised world wants development to stop, most certainly not in their own countries and not in the developing countries either. But if development conflicts with the sustainability of the environment then it must take a back seat. In the developing countries the priorities are, understandably, reversed. Having thoroughly polluted the planet during their relentless drive for industrialisation, the rich have no right to demand that the poor pay heed to it at the cost of jobs and productivity from the very start. A much saner course would be, for the rich, to reduce their demands on the environment to accommodate the inevitable growth in the demands of the poor, at least until such time as the pressure of poverty on the latter eases.

This, unfortunately, is where we move out of the realm of morality and enter that of economic power politics. The rich are prepared to spend any amount of money to reduce pollution and environmental damage, in their own countries, but reducing their demands on the eco-system. This pressure has a genuine moral base, but is propelled by the hard edge of self-interest. Cleaning up the environment is expensive and pushes up the cost of production in the rich countries at a time when their high labour costs are already pushing more and more global manufacturing into the developing world. If the latter are allowed to get away with neglecting the environment then their cost advantage will only increase. They must, therefore, be forced to observe the same environmental standards as the advanced countries. Bit by bit the developing countries have conceded implicitly that the rich have the right to continue enjoying a lifestyle that requires the appropriation of 34 per cent of the world's natural resources by six per cent of the world's population (to quote the figures for the U.S.) and that the developing countries must bear from the start the responsibility for protecting the ozone layer, preventing global warming, protecting the biological diversity of their natural forests and grasslands and preventing further pollution in the course of industrialisation.

The poor argue that since the rich will benefit from the measures that they are asking them to take, for instance by not cutting down rain forests, and cleaning up flu gas and vehicular exhaust emissions, they must bear most of the cost. The rich nations are inclined to accept this argument in principle, but there is a wide chasm between what the developing countries say they will need and what the industrialised countries are prepared to pay. How environmental costs will be shared, and whether the transfers to the developing countries will be through grants or loans are to be resovled.

The developing countries will receive only a fraction of the additional cost that they will morally bind themselves incurring, and a good part of that will come as loans which will only postpone the charge on their scarce capital resources. The conflict between development and environment will therefore continue, and inevitably, the latter will continue to be sacrificed for the former.

The root cause is the use of prodigious, and still growing amounts of energy per person, and the fact that these concentrated doses of energy have so far only been obtained from fossil fuels. The carbon in these fuels was trapped by millions of years of photo-synthesis by the earth's then abundant vegetation, but is being released back into the atmosphere at a rate that will exhaust the stock in as little as another hundred years.

If development is to be reconciled with the protection of the environment, then the starting point must be to find concentrated sources of energy that are environmentally benign. The obvious answer is to shift from fossil to renewable sources of energy. Only then can the conflict inherent in the very concept of sustainable development be resolved. What is more, since the main source of renewable energy is the sun, it is one that the developing countries are particularly well positioned to exploit.

The advantages of solar thermal and biomass energy are too many to state. Both technologies are already at that crucial point in their development where if the true social cost of conventional coal-based power generation were to be calculated, including the cost of mining the coal and transporting it to the power stations, the cost of building, of expanding new be withdrawn from other uses to dump the waste flyash upon and the cost of cleaning up the flu gases from the power stations, both are very close to being economically viable. The need is to adopt taxation and pricing policies in the rich and poor nations that would, with the minimum of intervnetion in the play of market forces, tilt the scales in favour of further developing and using solar and other forms of renewable energy. This priceless opportunity should not be lost.

GLOBAL PARTNERSHIP FOR SAFE EARTH, ENVIRONMENT AND DEVELOPMENT

The removal of poverty is an integral part of the goal of an environmental strategy for the world. The concepts of inter-relatedness, of a shared planet, of global citizenship and of 'spaceship earth' cannot be restricted to environmental issues alone. They apply equally to the shared and inter-linked responsibilities of environment and development. The global community must address the issue of global development in order to move towards an environmentally sound planet. But the problem of poverty remains as serious today as it was two decades ago. It is this goal of removing poverty, of development at the global level, which must be addressed if environmental problems are to be tackled. We have a difficult task ahead. We have to achieve economic development with minimum damage to our environment, for this alone can sustain development. Unlike the powers that could exploit the vast resources of their colonies and industrialise rapidly, and pollute the global eco-system, the developing countries today are searching for more sustainable and environmentally friendly solutions.

But these solutions have a cost. History has led to vast inequalities, leaving three-fourths of the world's people living in less developed countries and one-fifth below the poverty line. The long-term impact of past industrialisation, exploitation and environmental damage cannot be wished away.

In the remaining part of this century and in the next it is only right that development be more conscious of its long-term impact. Our common future can only be achieved with a better understanding of our shared responsibility, for the present and the past.

At a global level, the common future depends on sustainable use of natural resources. The current pattern of development based on high resource consumption, high waste and low recycling, as practised in the North, is not sustainable. Yet it is in this direction that much of the South is heading.

The alternative paradigm of improving standards of living without inflicting heavy environmental damage poses its own problems. The developing countries would require access to technologies that would enable them to avoid the dirty industry route to development. Many of the technologies required are new, some are still being developed. The major costs of pollution control, and the limiting options from an environmental point of view are facts that the early developing countries in the North did not have to consider. However, the cost of these more stringent norms is the price of development today. These costs need to be shared if the developing countries of the South are to follow alternative path — a path that bypasses the route of severe environmental damage.

Sharing responsibility means sharing costs, rather than aid. Funding mechanisms that are essentially donor-driven need to be replaced by more open participatory mechanisms which involve developed and developing nations. It is only on this basis that we can approach the building of a common future.

Developing countries are equal partners in any process that discusses global issues like environment and development. Their inputs are essential in decision-making to bring in local perspectives and to prevent generalised solutions.

Evolving alternatives can only be done through a proper understanding of the local situation. These are best done where decision-making is not centralised. Environmental standards and mechanisms to deal with environmental issues also need to be worked out on a country by country basis, if not in a more decentralised way. An international funding mechanism or a trade policy should not become the instrument to impose environmental conditionalities.

The countries of the South are among the richest in terms of biodiversity. This biodiversity is the source of a number of products that today benefit all humankind. Advances in biotechnology have increased the range of products of pharmaceuticals, agriculture etc. by modifying and combining the genetic resources collected largely from the South. With much greater financial and research capabilities in the North, these new products are being made available through companies of the developed world.

Issues of biodiversity must be viewed along with the sharing of biotechnology, sharing of the results of R&D and the commercial benefits derived from the biomaterials. (For instance as stated elsewhere, the more than 50,000 varieties of rich found in India have provided a tremendous base for research of rice.) One recognises that biotechnology research must be adequately compensated. But mechanisms other than the presently prevalent system of intellectual property rights are required. The researcher has to be compensated without restricting use as implied in IPR.

Access to environmentally friendly technologies also needs to be easier and much more widespread. In many cases, existing technologies would need to be replaced. The cost of the newer technologies and of pollution control mechanisms is likely to be higher than that of the older processes. However it is important that the wrong decision is not taken due to short term financial constraints faced by developing countries. Mechanisms would therefore have to be found to subsidise the change-over.

India shares the concern about ozone depletion and understands the need for coordinated action. But it also recognises that the problem has essentially been created by the developing countries who must take principal responsibility for it. A reduction in emission levels of CFCs in the North should not, however, imply a ban in the South on even the small levels of CFC use — use that arises from refrigeration and other requirements which are, if anything more necessary in the much hotter tropical countries of the South. It is therefore necessary to think in terms of quicker and less costly methods of transferring the newly developing alternative technologies to the South.

The most significant aspect of the climate change convention at the Earth Summit is the tacit commitment by the developed countries to restructure the Global Environment facility (GEF, the global fund for environment) to make it more equitable and transparent in its administration.

The developed countries also stand committed, to meeting the full incremental costs incurred by the developing countries in switching over to new environment-friendly technologies.

The climate change convention as it was agreed upon and finalised makes it amply clear that the commitment of the developing countries will only be to the extent of cooperation in financing and technology transfer on the part of the affluent countries.

Apparently no time-frame for the reduction of greenhouse gases has been worked out. The lack of a time frame is bound to meet with disapproval from aggressive environment movements in the West which see that the global warming problem is always ready with us and that any delay could defeat the very purpose of a climate change convention. The quantum of greenhouse gases already emitted into the atmosphere may have already committed the world to a warming of about four degrees celsius over the next 50 to 100 years.

The points made forceful by the developing countries and accepted by the climate change convention document are the overriding priority of the developing countries for economic development and poverty alleviation programmes, the need for new and additional financial resources for the new conventions being worked out, the need for restructuring the GEF to make it more transparent and equitable (to remove donor bias), the recognition that the GEF itself may only be an interim financial instrument, the recognition that full and incremental costs for switching technologies must be paid to the developing countries, and finally although the responsibility for the state of the environment was shared by all countries, it was 'differentiated'. This vitually recognises the fact that the developed countries have the much greater responsibility since they have caused the problem.

Today it may not be possible for governments to ignore the grassroots environment movement in their own countries. The movement has been gathering strength over the last ten years and in some countries 'the Greens' are not politically insignificant. Recently the scientific evidence has also strengthened the worst fears of environmentalists, and the doomsday predictions of a warmer planet with rising sea levels appear dangerously close to what has been measured and observed on the ground by scientists.

The next big thing is bio-diversity. The developed countries have to commit substantial funds for the GEF; that would be the most ticklish issue.

GLOBAL APPROACH

Global Fund to Green the Earth

The world has 27.6 per cent of its land under forest cover and this has to be increased by 2.4 per cent over the next eight years to meet the target of greening atleast 30% of the earth by 2000 years.

Industrialised world is responsible for 80 per cent of the pollution. An equitable global package has to be evolved in a more supportive economic environment where there is a sharing of control and a more democratic structure.

A global package with the global programme of action, a specific decision on additional financial resources, a decision on technology transfer at preferential rates and an inter-governmental institutional structure to monitor the follow up are needed.

Equal Commitment

The international community will need to formulate programmes that can promote and finance the transfer of sound technology without any impediment caused by intellectual property rights. Developing countries cannot meet the environmental challenges effectively if they are deprived of the access to technological alternatives.

An unjust international economic order cannot be replaced by an equally unjust international environmental order. The developing world and legitimate aspirations to achieve lifestyles of comparable security as that of the developed world, but the developing nations are committed to achieving that standard of living without following the ruthless route.

A new global partnership would not be based on unilateral action but on common concern, based on equal weightage and common concerns, no distinction is to be made between environment and development, between global and national concerns; a massive programme of regreening rather than just the containment of pollution; a stronger role in dealing with these mechanisms, and the economic and social issues arising out of these linkages to be made transparent and accountable. A distinct and separate mechanism is necessary to ensure the transfer of new and additonal financial resources and environmentally friendly technology on non-commercial terms.

Unless there is a perceptible change in the attitude of the richer countries brought about by the badly needed heart-searching, the re-affirmation of the target at the Rio meet would not go very far. Third world countries such as India, Brazil and Indonesia could set an example by formulating new policies, plans and strategies for sustainable development and poverty elimination. There is no doubt that "poverty is the worst polluter." We have to implement faithfully the new plans, for which, to begin with about a Budget of about Rs. 500 crores (i.e. about 0.07 per cent of India's GNP) can be provided. Gradually the budget could be increased to about 0.5 per cent of the national GNP by the end of this decade to clean up man-made mess.

We do not need any foreign help to start a massive programme of reforestation and planting of trees, to enforce rigidly the laws relating to emission of CO_2 (and carbon monoxide) from the vehicles on our roads. Our priority number one is to stabilise our population growth. Do we need foreign aid to clean up our rivers, lakes and waterways into which industrial effluents, sewage and so on are thoughtlessly pumped in large quantities? We should mount an R & D effort to find environment-friendly substitute for CFCs that give rise to ozone layer depletion so that by the end of this decade, we can use such substitute.

'Polluter Pays' principle states that the polluter should bear the cost of measures to reduce pollution to ensure that the environment is in an acceptable state.

The biggest global polluters, should contribute one per cent of their Gross National Product towards a Global Environment Facility to enable steps to be taken worldwide to reduce carbon dioxide and other greenhouse gas emissions and to undertake other measures to conserve biodiversity and forests. The one per cent annual contribution would have amounted to $125 million each year. With the developing countries expected to find another $600 million every year to invest in the policies and programmes leading to a better environment and tackling poverty eradication problems that are linked to it.

The richest countries which have been playing the lead role in exhausting and exploiting the global environment should pay for it.

They could raise the resources by cutting down on their highly inflated defence budgets or by imposing a sort of environment tax based on emission of carbon equivalent.

The developing countries are simply offering to cooperate in taking up the global task of stabilising and then reducing global greenhouse emissions, a task which is essentially that of cleaning up the environment mess created by the wasteful and consumptive lifestyle of the rich countries.

It is perhaps worth repeating the scientifically well-established statistics that of a total of 5 billion tons of carbon equivalent emitted each year into the atmosphere by human economic activity, nearly three-fourths is being emitted by the top industrialised countries.

PART III

APPENDIX
(Some More Issues Calling for Immediate Attention)

I

ECOSYSTEMS AND THEIR PROTECTION

Evolution Profile of Eco-Systems

New discoveries have thrown an entirely new light about the origin of life on earth.

Traditionally, deep ocean seafloors have been regarded as intensely dark, cold and lifeless deserts. The new discoveries have shown that this is far from being the case. The supposed deserts have been found by oceanographers to support mineral-rick oases thriving with self-sufficient communities of marine animals, the crucial difference between these communities and other living organisms on the planet is that they flourish on energy drawn from the earth's interior and not from the sun.

All other marine animals live according to the same fundamental principles as life on land. Their survival is based upon food supply generated by plant growth, fuelled by energy from the sun built with the same basic elements — carbon, nitrogen, and hydrogen. The plants are browsed by herbivores as on land, and these in turn become food for the carnivores.

There is, however, one great difference between life ashore and life at sea. Virtually all oceanic plant life occurs as single cells or simple chains of cells so minute as to be visible only microscopically, so that the herbivorous marine animals have to industriously filter the food from water in which it is suspended before consumption. Only along the margins of the oceans, on the shoreline or just below it, are there large enough plants comparable to land plants. While it is true that this vegetation is a food source for numerous marine animals. It is relatively un-important in this capacity in the oceans as a whole. Marine plants reach their maximum size in the giant kelp (large brown seaweed) beds.

The geologic events of the earth's surface are at the mercy of plate tectonics, the driving force that changes it. The rocky crust of the surface is broken mainly into 20 massive slabs called lithospheric plates, each composed of continental rocks and ocean floor basalt (igneous rock or rock produced by volcanic activity), and a portion of the upper mantle (the part of the earth between the outer crust and the core. The cold, rigid plates ride upon the top of the partly molten asthenosphere, the layer immediately beneath it and composed of magma and crystallised rock.

As the plates so ride, they respond to heat rising from the earth's interior and move, shifting continents and ocean basins around the globe. Plate tectonics has moved continents or parts of them around the planet for thousands of millions of years.

Where the plates pull apart, new seafloor is created at these rifts or "spreading centres." Magma from beneath the crust wells up to fill the gap. Magma is molten rock, the material from which all igneous rock crystallises. Ignoring how exactly magmas originate, a body of magma will be at high temperature and high pressure when it is within the earth. As it moves up towards the surface, it cools losing heat to the rocks around it, until it is cool enough to crystallise and form igneous rock.

Magma contains silica and a number of major elements (aluminium, iron, manganese, calcium, and many more) and volatiles (water, chlorine, hydrogen, oxygen, nitrogen, etc.). As it rises the volatiles are released as gases (carbon dioxide, hydrogen sulphide etc.) when the pressure on the magma grows less. Thus when the magma actually reaches the earth's surface, it consists of a mixture of lava (molten rock) and gases, and gives rise to volcanic eruptions.

As the plates seaprate, lava wells up from below, flows on to the seafloor, cools in the frigid water, and bonds to older basalt, creating a new ocean crust with cracks and fissures. Now follow huge and complex hydrothermal processes in this rift system, when ocean water begins to circulate in it. How for down the water descends remains unknown: but it is thought to be several kilometres. The intense heat and high pressure from the magma below causes the water to leach (remove away and out) minerals from the surrounding rocks. The water is forced back up to the seafloor by convection and erupts through vents as scalding jets, at 300°C or more, or in warm streams. As it strikes the cold sea water, the minerals (iron, copper, zinc, etc.) precipitate and build up substantial sediments.

The entire content of the earth's oceans circulates thus once in every 10 million years, profoundly affecting the chemistry of the seas.

Movements along the spreading zones crack and tilt the new blocks of crust. The spine of rifts and ridges called the Mid-Ocean Ridge is the longest mountain range in the world, and winds 74,000 km round the globe. The first discoveries of deep ocean marine communities were made along it. More recently, other discoveries have shown that there may be more deep ocean ecosystems than was ever thought likely after the initial startling finds.

As new seafloor is being formed, equally, old ocean crusts sink back into the earth's interior, creating deep ocean trenches. Thus ocean crust is being continually created and destroyed: this has to be so because the earth is not expanding. Earthquakes sometimes accompany plate motion and crust action, and are therefore concentrated along ridges and trenches. Thus, the lines of earthquake action are well-known.

Early investigations on seafloor spreading had predicted that there might be hydrothermal vents. But not much life, if any, was expected at such frigid and sunless depths, this was a reasonable assumption, as devoid of sunlight, the acknowledged energy source for plants to produce organic compounds crucial to sustaining all life on land and sea, living organisms were not expected to thrive. Investigators were astouned to find a varied and abundant life surrounding the vents.

How did the communities of marine animals find the energy which drove the global hydrothermal circualtion system at the spreading centres. Fluids escaping from the vents contain high levels of hydrogen sulphide. This gas, as we know, is toxic to most life. Nevertheless, certain bacteria can metabolise (build up into living matter) it. Multiplying rapidly, there bacteria form the basis of the vent community food chains. Higher organisms feed on this bacterial profusion or on other creatures that do. Some, such as tubeworms and clams, mountain symbiotic (by attachment or living one within the other) relationships with bacteria that live within their bodies and provide them with a steady food source.

Beacause of the dynamic geological processes at work, vents last only a few tens of years. When they die out, the marine animal colonies around them die out too, but already larvae from these animals may have been dispersed over great distances to colonise new vents. The profusion of closely related species at isolated vent sites suggests that this may be happening. Other communities may also evolve separately, for which there is some evidence. Recent discoveries of a site swarming with eyeless shrimp (a familiar ten-footed crustacean) confirmed the possibility of ever more and varied ecosystems at spreading centres.

Vents at depths 2,500 m and more have revealed astonishing densities of life around them. Concentrations were as much as 500 times more than that normally found in deep ocean. Crevices in the lava were filled with rich colonies of tubeworms, limpets, mussels, shrimps, clams and crabs, many of them previously unknown species. Less than one per cent of the global spreading centres and systems has been explored: a fact signifying that this biological realm of a world without sun has yet been so little investigated.

The earth's crust is not of uniform thickness, it varies from 10 and 75 km and under the continents to around 8 km under the oceans. The top of the oceanic crust is generally 3 to 4 km lower than the continental crust. There are other differences between the continental and oceanic crusts, among which is the topographic features of the oceanic ridge, which is where current volcanic activity is concentrated and new oceanic crust is currently being generated. The ridges rise 2 to 3 km above the ocean floor, and are the locations of activity for seafloor marine life.

Extreme and elaborate adaptation to the conditions at the vents are the keynotes to the survival of the organisms. When better understood, new swathes of knowledge will be added to extraordinary adaptive evolution and refinement. The existence of such complex biological ecosystems raises new and exciting questions about the origin of life on earth.

The established belief has been that life on earth originated in a sunpowered environment. Now arguments are beginning to be put forward that geothermal energy at deep ocean vents may have allowed the first primitive organisms to evolve. This virtually new field of research will no doubt provide heated debate in the coming years on whether life originated at the vents or only adapted to them.

Vast new horizons of investigations have been opened up for

oceanographers in the chemical, physical, geological, and biological disciplines. *To what extent these sunless ecosystems will be affected by the massive pollutions human activity is today pouring into seas is an exciting subject to study.*

OCEANS AND THE WATER PLANET-EARTH

The ocean covers atleast seven-tenths of the surface of our planet and yet most of us tend to see oceans only as vast expanses of salt water, as barriers between land masses. In reality, it is the other way round. There is a world ocean with land masses as barriers. When sailors boast of having sailed the seven seas, they have really traversed a single ocean. Like the atmosphere, the oceans know no physical boundaries.

The oceans not only physically cover a large part of the land, they are more than just water. With their rich cache of fisheries, minerals and potential energy, they are an invaluable resource realm of our planet. It is the oceans that decree the climatic conditions of the lands. Oceans act as a gigantic weather machine distributing heat and cold through the currents, controlling winds, rain, air pressures and humidity. While the temperature may reach +136 degree F in the Sahara and drop to -126^0C in Antactica, the temperature of the seas rarely exceeds +80 degree F or drops to +30 degree F. This is because water is a much better heat conductor than air or earth and it requires huge quantities of heat to evaporate, or to freeze. All this makes the ocean the great climatic moderator of planet earth. The ocean is the cradle of creation.

The long and complex process of evolution of life began below the surface of the sea. Oceans provide directly to life forms, the water they need for survival and growth. It is water that is loaded with vital salts, dissolved gases and minerals. The comparative constancy of temperature enables many species to thrive without having to adapt to wide variations in temperature. And even today life on land is dependent on life in the sea.

Ocean ecosystems are, in reality, continuous. This huge single ecosystem has several distinct regions which have distinguished differences in depth of water, its temperature, saltiness and the availability of nutrients. Permutations and combinations of all these, in turn, provide habitats for a variety of plants and animals.

The ocean is a huge and complex living ecological machine in which every part interplays with every other. This ecosystem is a vital and integral part of our planet. A dead sea means a dead planet.

THE CORAL ISLANDS IN THE SEA BOTTOM

Six hundred thousand square kilometres of forests in the Sea? Indeed! These are the world's coral reefs—marine equivalents of tropical rainforests. The resemblance is more than just in the way they look — groves of corals of different kinds with trunks and brances, just as trees in a forest. A coral reef teems with life. The life of the dense and diverse community of organisms that live in, or feed in, the thickets and branches of the reef. Almost every crack and crevice of this watery jungle is occupied by some marine creature.

Coral reefs have existed for close to 500 million years, making them one of the earth's oldest ecosystems. They flourish in tropical waters with temperatures between 25^{0}C and 29^{0}C. As they need the right temperature and sufficient light to grow, reefs flourish in shallow waters.

The creation of a reef can take centuries. Reefs are built up of millions of minute animals called coral polyps, each of which lives inside a protective skeleton of limestone. It is the limestone that forms the main substance of the reef.

OCEAN POLLUTION

Since the dawn of time, the oceans have been the ultimate receptacle of all waste. Fumes and dust from volcanoes and sandstorms, dead flesh and bones, rotting trees, tonnes of silt from fields and rivers — they have all been enveloped into the fathomless waters. And these "wastes" were transformed and reused as fuel for the machinery of Nature. For million of years, we thought of all these merely as "a drop in the ocean."

Most people believe even today that the ocean is too vast to be damaged by man's activities. If the oceans have worked as gigantic 'waste treatment works' for millions of years, why can't they do so forever? The difference is not that the nature of the oceans has changed, but that the nature of what enters this enormous self-purifying filter is no longer the same.

We tend to think of ocean pollution only as waste being thrown overboard by ships at sea or accidental oil spills from huge tankers. But at least *85 per cent of pollution of the oceans has its source in human activities on land. These activities include agriculture, industry, urbanisations and its resulting outputs.*

In fact, what are we pumping into our oceans, and where does it all come from? *From Fields: Pesticides and herbicides used in fields find their way through the washed away silt and irrigation water into rivers,* and eventually to the ocean. These chemicals are *not only not biodegradable,* they also pass through the marine food chains, getting concentrated at every stage. *Thus they pose a threat not only to the tiny marine organisms, but even to the fish that feed on these, and sometimes to the humans that eat the contaminated fish.*

This was tragically illustrated in the early 1950's *when seafood contaminated with mercury from a local factory killed over 400 people* at Minamata Bay in Japan. By 1990, almost 40 years later, nearly 3000 people had been diagnosed *as suffering brain damage from mercury poisoning which could be traced back to Minamata.*

From Factories: Effluents from industry reach the oceans through rivers and even from the air. Industrial smoke will be carried by wind only to fall back into the ocean as rain or snow. *Heavy metals such as mercury, cadmium and lead, liquid chemicals* of all kinds are discharged into rivers which flow to the seas.

From Oil Spills: Oil in the oceans becomes new wherever there is an oil spill or leak from an oil tanker. While such spills do have disastrous consequences for the ocean waters and their dwellers, they seem to overshadow the millions of tonnes oil that daily enters the oceans. This is crude oil that is discharged from navy and merchant ships all over the world and from routine cleaning of oil tankers on high seas. These ships wash out their tanks before taking on new cargoes. Every year over 1.1 million metric tonnes of oil is deliberately discharged from tankers.

One of the worst oil spills in history was an outcome of the Gulf War. Around 6 million barrels of crude oil were released into the Gulf. Most of this washed up along the cost of Saudi Arabia — with devastating consequences. Over 30,000 sea birds were killed. Mangroves and salt marshes in the area were destroyed. Scientists believe that it will take more than ten years for the *ecosystems of the area to recover.*

Every year about 4 million tonnes of *oil contaminates the sea.* Only half comes from ships — the rest originates from the land.

From Plastics: While oil in the waters can cause havoc, oil may in time break down, and in the long run, the environment may recover. But plastic in the waters does not ever break down. And it is more than just a floating eyesore. Turtles and fish suffocate from plastic they have swallowed mistaking it for something edible.

A PROFILE OF EARTH

What is inside of our planet? The deepest hole drilled so far is 12 km. nothing compared to the 6.370 km radius of earth. Measurements are through studies of seismic waves and of the Earth's magnetic field.

Structure of Earth

The shockwaves generated by an earthquake or a large nuclear explosion are of two kinds: compression (P) waves propagate by the alternate compression and refraction of the Earth material, while in the shearwaves the vibration of particles is perpendicular to the direction the wave. By measuring the variation of the speed of travel of these waves as they traverse through different regions of the interior of the Earth, seismologists have found that the earth has a shell structure rather like coconut. There is an outer crust 30 km thick but under the ocean floors this crust could be as thin as 10 km. Under this crust is the mantle. The outer part of ths mantle is solid, about 1,000 km thick and the inner part is viscous about 1,900 km deep. Within the mantle, insulated by a layer of oxidised metal is molten iron and nickel at temperatures of between 2,000° C. At the centre of Earth and within the liquid core is a solid seed with 1.220 km radius.

Oldham found that S-waves which were received 120°C from the epicentre of the earthquake were delayed by about ten minutes. The waves must have passed through a core which slowed them down. It is also known that the pull of the Sun and Moon deform the Earth by a few tens of centimetres. This deformation must be due to a liquid inside and not a rigid solid core. This liquid core also refracts the seismic waves away from the angular region causing as shadow zone for them. In 1914 Gutenberg measured this shadow zone for them to be 105°C to 140°C and so computed the radius of the core to be around 3,500 km. Extremely sensitive seismometers were brought into operation at centres round the globe. They detected certain seismic waves of low frequency which resonated in the Earth, rather like the tones of a bell. This behivour could be understood only if a solid seed of about, 1,200 km radius was located at the centre of the Earth.

As regards the composition of the core, all evidence suggests that Earth was born out of the primeval gas and dust. As this material collapsed under the influence of gravity, the gravitational energy converted to heat and fused the material into our planet. This heat plus that due to the distintegration of radioactive elements cause convection currents in the liquid core. The heat then escapes out through the mantle and the crust into outer space. Models based on the characteristics of the material of the core estimate the temperature of the seed to be around 5,000°C, ten times less than at the centre of the Sun.

Born out of the same dust as the Sun, meteorites and the other planets, the Earth should be of the same mixture of elements as all of them with an iron to silicon ratio of 0.9. But mineralogical measurements of samples from the mantle indicate that they have very low iron content. Therefore, the molten core and the seed should be made of more or less pure iron with perhaps four per cent of nickel. Other light elements, mainly sulphur and oxygen also exist but these migrate out to form an oxide layer between the core and the mantle. It is very likely that the solid seed is a large crystal of iron, or perhaps an alloy of nickel and iron, growing slowly as the Earth cools. Under the tremendous pressures inside the Earth, the crystal is far more dense than the alpha iron phase that is common and it is suggested that is beta-iron with hexagonal close packed crystal structure.

Earth as a Magnet

The Earth's magnetic field serves a vital purpose by protecting life on the planet from the lethal atomic particles emitted by nuclear reactions going on in the Sun. Contrary to popular belief, the iron in the Earth cannot be a permanent magnet as the magnetisation of iron disappears above 77°C, the Curie point. But the liquid and electrically conducting iron core rotates due to convection currents as well as turbulence encouraged by the continuous crystallisation of the seed. This rotation generates a magnetic field in the much of the same way as an electrical dynamo. Measurements on the variation with time and the geographical distribution of the magnetic field show that the liquid core is flattened at the poles, but the solid seed is elongated along the axis of rotation, the polar axis.

The connective movements of the core also has its side effects on the surface. It is known that the braking effect of the ocean tides gradually slows down the spin of the Earth, lengthening each day by about two seconds every ten thousand years. Further, there are irregular fluctuations in the length of the day due to the turbulent oscillations of the core. Perhaps long ago, these oscillations might have come into resonance with the attractive solar forces and caused major changes on the crust such as continental drifts.

Role of Spin

The spin of the Earth encourages the axis of rotation of the molten core to align itself in a north-south direction but it does not seem to matter whether the magnetic axis is parallel or antiparallel to the geographic axis. These days, Earth's magnetic field has its north magnetic pole in the extreme north while the south magnetic poles is in Antarctica. But this was not always so. Geologists have traced changes in the magnetic field by examining rocks that solidified at various times and so retained the direction of the field.

From these studies, it is known that every 2,50,000 years the field flips over and the north magnetic pole shifts to Antarctica. This reversal of course terribly confuses the birds and animals that depend on the magnetic field for their migration habits. The cause for this flip in the magnetic field is not clearly understood but recent evidence suggests that these magnetic somersaults are important. Till a few years ago, scientists believed that turbulence in the core broke up the correlated rotation of parts of the molten core. The magnetic field then split up into several disoriented fields and these gradually coalesced into a single one but upointing in the opposite direction.

From recent geological and rock magnetism studies, it is felt that the original field did not disintegrate but that the earth magnet slowly swung around by 190°C. The north magnetic pole travelled down the Pacific Ocean while the south pole went up through Africa and Europe. This migration route was followed even in earlier reversals.

The molten core being isotropic, changes in its rotation should not follow a particular migration route. The inner surface of a rigidly frozen mantle of the Earth must have an influence on the motions in the liquid core in the same way the rocks on a seashore affect the ocean waves. Perhaps 125 million years ago a large piece of the lower mantle might have broken off and as it travelled out towards the crust, halted the rotations and so affected the magnetic field. This hot piece of mantle, called a 'super plume' must have had its profound effects on the surface of the planet. Vast volcanic activity, pushing up the sea-floor, increasing the temperature of the earth by ten degrees with considerable emission of carbon dioxide, must have induced cataclysmic floods and devastation. After the flood, the warm and shallower seas plus the excess of carbon dioxide encouraged the beginning of life on Earth again. (Nataraja Sharma)

ENORMOUS DAMAGE AND DESTRUCTION BEING CAUSED TO THE EARTH AND HOW TO CREATE AWARENESS OF IT

The environmental degradation has reached a critical point, and if mankind does not take preventive action, the result will be irreparable and tragic. What we need is a capacity and willingness to question current notions of economic growth and rethink "our basic values and visions of progress".

Even more, the important objective is to realise the horrendous situation that we are in at present. The main problem here is the denial by vested interests — of the terrible ecological deterioration. Such a denial breeds a thinking which assumes that "we can get by with minor adjustments to business as usual".

What we need, then, is to realise that the problem exists, tackle it even though it would involve some sacrifices.

We cannot muddle through and take policy decisions that will define the way our children will live in the next century.

A rather grim state is happening to the biodiversity which refers to the ecosystems, species and genes that together make life on earth both pleasant and possible. 50,000 invertebrate species are becoming extinct every year due to the destruction of their tropical rain forest habitat.

While this is bad enough, worse is the loss of genetic diversity within species because this nullifies the work of farmers who have created genetic diversity over the last 2,000 years to increase agricultural production. Ryan estimates that the 30,000 different strains of rice planted by Indian farmers only a few decades ago will be reduced to a mere ten by the year 2005 if the present trend of destruction of rice gene pool continues, thanks to the development of high yielding varieties.

Nicholas Lerissen shows us the enormous extent to which nuclear waste has grown during the last two decades. While the waste is increasing at an exponential rate, there is yet no way to dispose it. "Like mirages in the desert, safe, permanent methods of isolating radioactive materials disappear under scrutiny as quickly as scientists propose them. What is really worrying is, that nuclear waste cannot be destroyed and that no amount of isolation can keep it from affecting the biosphere. For example, in India, less than 1 per cent of the enormous funds spent on nuclear power is spent on waste management.''

The deteriorating state of worker health, indiscriminate mining, urbanisation among others, are all examples of the way, mankind is misusing, not only the environment, but also a large section of mankind itself. (S. Ambirajan)

SOIL DEGRADATION AND CORRECTIVE MEASURES

What is soil? The simplest definition is that ''it is the medium in which plants will grow.'' Plants need an aqueous solution carrying minerals from which their roots can absorb food, and air so that the roots can respire. Soil is thus a complex of air, water, minerals derived from the weathering of rocks, organic matter derived from the break down of plant and animal debris and living organisms of bacteria, fungi, nematodes (worms of unsegmented cylindrical shape), and protozoa (unicellular and microscopic animals) that decompose all the organic matter that enters the soil.

Soil may also be alternatively defined as ''the surface layer of the mantle of rock waste in which the physical and chemical processes of weathering cooperate in intimate association with the biological process.'' All these processes depend on climate, and resulting soils vary with the climate in which they develop. But there are other factors, such as the nature of the parent rock or other deposit in which the soil has developed, the relief (contour) of the land, the length of time during which its development has been in progress, the ''age of the soil'' mass and character of the vegetation growing on it, and the superimposed effects of cultivation or other human activity.

Soils support the plants on whose existence human life depends. A little realised fact is that soils have determined the course of human civilisations from the beginnings of history. An inch of soil takes some 200 years to lay down (but can be swept away in a very few years). A natural profile of soil formation and development is as follows: at the bottom is the bedrock or parent rock material; above this is weathered rock mixed with organic matter washed down from the surface; and above this is the topsoil, which is the ''living soil'', the seat of fertility. Soils may take thousands of years to reach a state of equilibrium, a state of layered arrangement.

Soils, viewed on a human timescale, are a fragile resource. They are a renewable resource only when viewed on timespan of thousands of years. Of the entire global land surface, only about 20 per cent is any good for grassland agriculture and a further 11 per cent for cereal production. Present estimates suggest that 0.3 ha. of arable land is available per head of world population, but when human numbers rise to six billion by the end of this century, that figure will fall by a sixth, to 0.25 ha. If food consumption comparable to western standards is to be enjoyed by the rest of the world, 0.5 to 0.7 ha. of arable land per head is required. So, even with a standstill population, the bulk of the world's peoples are at a disadvantage.

It thus becomes obvious that the *current disregard for soil resources is dangerously misplaced.* All fertile soils need to be cossetted, and all fringe soils only put to use that is consistent with conservation principles. What is actually happening, however, is that there is a *large-scale erosion and pollution of this most vital resource from uncontrolled human activities.* Worldwide, an estimated 20 billion tonnes of topsoil are washed or blown off croplands every year. And, more than three billion hectares are at risk of erosion.

Natural erosion occurs mainly from water and wind, and has gone on from the world's beginnings. Natural processes over which man has no control have created erosion but have also formed fertile land, so that the two have worked out into some sort of balance. Let us look, briefly, at how natural erosion occurs before considering human-agency erosion.

Water erosion is minimal where the soil has a strong cover of vegetation, particularly if it is bound together by a mat of interlacing grass and plant roots. But where the soil is poorly constituted and without this reinforcement, the result is very different. The best of the topsoil is easily moved to lower levels and into rivers by rainwash and rainsplash (the disastrous effect of the latter has been discovered only in recent years). The runoff forms a thin sheet of muddy water on even ground or makes rills where the land is crossed by rifts and gullies. The topsoil is carried away and lost for ever. Then it is the turn of the next layer, and inferior soil, long before it has had the time to be replenished by the slow weathering of the subsoil. Within a few years, the soil as a whole will be lost either by being washed away or becoming loose and blown away by the wind, the second agent of natural erosion.

Where the moved topsoil is laid down again as a result of the relief of the land or other reasons, fertile land may develop.

But, when human use of the soil for growing crops proceeds without any regard for long-term effect, both the natural agents, water and wind cause destruction. Widespread and expensive reclamation measures have, partially conserved and improved what was left of the once fertile soils. Similar devastation occurred in the semi-arid soils.

Looking at a plot of fertile soil, who can imagine its astonishing range of benign qualities? Most of the soil's organic matter and living mocroflora are located here. *A single hectare of topsoil can store 1.2 million litres of water, grow enough grain to feed an average of 25 people, and filter between one and 10 million litres of rainfall each year.* The smaller members of the soil microflora, the bacteria, fungi, nematodes, and protozoa, known together as the soil 'biomass,' process all the plant and animal organic debris entering the soil. In doing so, they form the organic matter of the soil, recycle vital nutrients for use by plants, and return carbon dioxide to the atmosphere. And we are scarcely even aware of any of these facts. A soil without biomass would very quickly become infertile.

Besides serving as the medium for growing our basic foods, we ask the soil to do a lot of other things besides. *It has to store and filter much of our drinking water, receive and store wastes of all descriptions, suffer our contamination of it with toxic substances and radioactivity, and accept erosion and still grow crops as before.* If these deleterious processes proceed unchecked, the soil's answer will be a resounding 'no' to food production. *Soil degradation is one of the world's greatest environmental problems.* Yet it does not receive anything like as much attention as many of the difficult questions.

Even where soils do not suffer from water and wind, *following man's denudation of its topsoil,* they suffer significantly when subjected to long-term cultivation without appropriate safeguards. The economic pressure to generate export earnings has been transmitted to our natural resource base. As a result, soil erosion today can be described as epidemic in proportion.

Some soils are more prone to erosion than other when under continuous cultivation. Research has shown that the crucial factors are the extent of vegetative cover and amount of organic matter in the soil. Soil organic matter is the 'glue' that binds the soil particles together. Soil nitrogen is a good indicator of the presence of organic matter, and can be measured. Under constant cultivation, the percentage of soil nitrogen falls steadily because the rate of organic matter break down is increased and the rate of formation of new organic matter is reduced. The latter happens *because in removing the crops for our use, we reduce the returns of carbon and nitrogen to the soil.* These two vital elements are needed by the soil biomass to synthesise, or build up, organic matter.

In these processes, tropical soils are at greater risk because nitrogen, and therefore organic matter levels tend to be lower, the warmer the average soil temperature. Vegetative cover protection then becomes even more important to shield the topsoil against wind and rain erosion. Measurements made in tropical rain forests have shown that where the forests have been cleared, erosion from rain increased as much as a hundred times. Even with partial vegetative cover, the damage was much less. Slash-and-burn agricultrue was seen to be very destructive of the original vegetation and results in massive erosion. Once the nutrient residues of this vegetation are gone, the fertility of the soil is also gone. If crops are attempted to be grown they fail.

It is easy to see, then, that agricultural practices must be such as to avoid the worst effects of erosion, and why forests and other vegetative cover need to be retained to hold the soil together. *The disastrous effects of tree felling in the foothills are well-known. The beneficial effects of the restoration of the trees are no less well-known.* Thus, not only must agricultural activity be limited to more suitable soils but also there has to be a check on removal of vegetative cover and a need to return crop residues to the soil.

Soil degradation is also caused by a range of other human activities. *High among these is pollution.* Even if the soil remains in place, *fertility can be severely damaged by pollution.* The growing practice of using the soil as *a natural repository for all kinds of waste including hazardous and toxic wastes, has been insidious.* Practitioners of this attempt to lull fears by pointing to apparent ability of soils to absorb these wastes without showing ill-effects. *But this has been exposed as crude propaganda by examination of affected soils. The changes to the soils are very significant.*

Soil bacteria are of all sizes and shapes. Diversity is one of their great strengths. Some bacteria can fix (increase input) nitrogen, some can decompose organic matter, and others do many more things. Even such a practice as spreading sewage sludge on soils, once considered benign has been shown to be very harmful. Sewage contains heavy metals such as copper, cadmium and zinc. A striking experiment carried out on two adjoining tracts of soil, one which had received sewage sludge and the other was free of it, revealed that although the sewage-treated soil still contained bacteria, they had lost the ability to fix nitrogen. *What had happened was that bacteria that had this property had been killed, while those which did not, survived.* Thus no leguminous crops could be grown on the treated soil.

Sewage also contains large quantities of nitrates and phosphates. Excessive use of fertilizers (nitrates and phosphates) is justified by the argument that such inputs are the quickest way of increasing cereal yields. *But the overkill results in over-fertilisation which in turn leads to europhication (killing off animal-life by oxygen deprivation) from excessive algal growth and to draining of the nitrates into rivers and lakes to cause deterioration in fresh water quality.*

Soils can be adversely affected by other *forms of pollution: acid rain which eventually leads to leaching of aluminium into the soil and rendering it toxic to most forms of life, outletting toxic wastes from chemical, metal, and pharmaceutical industries, dumping nuclear wastes etc. Vast land tracts have become sterile — and extremely dangerous — from outletting nuclear wastes.*

The environment perspective of the developing countries is clear. The main global problem confronting the world is the rising emissions of carbon dioxide and other greenhouse gases threatening a vast climate change that could affect food production and create acute fresh water shortages. The Third World pointed out that the affluent nations had caused the problem. *The 20 per cent of the population in the countries of the industrialised North together emit 80 per cent of the global greenhouse gases.* 'The polluter must pay,' principle must be applied, and if any corrective measures are to be taken, they must give the money and the technology to make possible the changes.

VOLCANO AND GREENHOUSE HAZARDS

The eruption of Volcano is expected to chill the Earth slightly, reversing a trend of gradual global warming caused by the so-called greenhouse effect and the weather system.

Sulphuric acid particles blasted into the upper atmosphere by the massive volcano can be expected to decrease global average air temperatures by as much as one-half degree celsius. The sulphuric acid particles reflect solar radiation.

Based on atmospheric computer models and satellite observations, the predicted cooling effect is estimated to be about twice as powerful as the warming effect caused by the growing concentration of carbon dioxide and other greenhouse gases entering the atmosphere from the burning of fossil fuels.

"We will see a temprorary reversal of the global warming trend that we have had for the last couple of decades." The next two years will be noticeably cooler on a global average than recent years.

"A half degree of cooling, roughly speaking, would knock the probabilities for a warmer than normal season back to what they were 30 years ago. But it doesn't guarantee that it's going to be cool at a given place on the globe."

The cooling effect of the volcano is also expected to overwhelm the warming caused by the current weather phenomenon known as "El Nino". The "El Nino" even occurring every three to seven years, involves an abnormal warming of equatorial waters, followed by a warming of the atmosphere. El Nino's influence can shatter normal weather patterns over much of the Earth.

Recent measurements show that global surface temperatures did decline in, but rose again. Model simulations on preliminary 'Pinatubo' data, predict a dramatic global cooling. This is a case where the model is right and real world is wrong. "Global surface temperatures drop sharply in the next few months. We're going to have a hard time understanding why."

Alan Robock, a climate researcher also predicts a general cooling from Pinatubo, but pointed out that the effect is not uniform. For example, last winter was milder than expected in part of North America and in Europe as El Nino's wind brought a warming trend that overshadowed the effects of the volcano.

Satellite data indicate that the *volcano eruption lofted 15 to 20 million metric tons of sulphur dioxide into the upper atmosphere,* known as the stratosphere. *The sulphur dioxide gas joins with vapour in the atmosphere to form tiny droplets of sulphuric acid, which can remain aloft for several years of after an eruption.* By reflecting sunlight back into space, the droplets exert a cooling influence. But Pinatubo's volcanic aerosols will *gradually fall back to Earth as acid rain, and global warming is expected to resume as the concentration drops.*

"What goes up must come down". These aerosols are in fact falling out. So the volcanic eruption does not alter the expectation for long-term global warming. The eruption depressed global temperatures by as much as 0.3 degree celsius. The massive eruption chilled areas.

Robock's recent studies suggest that volcanic eruptions have been a principal cause of annual climate change in the Northern Hemisphere for several hundred years. "*The latitudes farthest from the equator experience the most cooling, and this cooling is most dramatic during the winter.*" The researchers reported disturbing evidence that Pinatubo's *sulphur aerosols are causing a 15 to 25 per cent depletion of the ozone layer at high latitudes near the poles, as well as indications of a drop in ozone in the tropics and possible at temperature latitudes.* The ozone layer protects Earth's inhabitants from the damaging ultraviolet radiation from the Sun.

SAVING OUR PLANET BY VERY SIMPLE METHODS

The need of the hour is to take immediate action to plug the widening gap in the ozone zone, and increase the forest cover. Financial viability to implement programmes is needed to save this planet from nemesis.

A few noble 'dos and don'ts' towards reducing, if not eliminating the pollution which is slowly turning our planet into a hell hole, are:

— The first suggestion is very simple. It exhorts us to keep away from automobiles to the extent possible. Any place within a kilometre can be reached on foot. And the bi-cycle can be the ideal means of transport when one has to cover longer distances, say up to six kilometres. By keeping away from automobiles as far as possible, we shall reduce the amount of exhaust from automobiles, which has been held responsible for much of the damage to our surroundings.

— The second step is to avoid wastage of water which is a scarce commodity, with increasing population, the demand on water is increasing though the supply has not been able to keep pace with the demand. So, water has to be utilised with care, for instance, ensure that the tap is closed the moment you don't need water. Avoid leaving the tap open while brushing teeth or having a date with the razor. One should realise how much water goes down the drain when the tap is callously let open while the brush brushes all over the teeth or the razor battled with the stubbles. Besides, in many households, taps leak. Prompt action to replace the washer shall save precious water. It shall also reduce the water bill.

— Thirdly, use paper products wherever possible. Shy away from plastics. Much of the bio-degradation has taken place due to plastics. Plastic bags and containers take many years to break into the natural elements compared to paper. They clog sewers, choke pipes, create complications whose magnitude has not yet been fully realised. While research is on to manufacture plastics with low lifespan, there has not been much breakthrough because of cost factors. That is where paper scores. Further, paper can be recycled. (Plastics too can be recycled. But paper scores over plastics due to the lesser danger it poses to environment.) So, it would be good to go back to normal paper packets and give up the ornate glossy plastic covers which are used to wrap whatever one buys.

There is also scope for saving energy if only we learn how to achieve it. Cooling of homes, even by air-conditioners will consume less energy if the windows have panelled glasses, with a double layer so that the targeted temperature is achieved quickly and retained over a longer duration. Other energy saving methods relate to saving cooking gas. Have all ingredients necessary for the dish ready, properly cut and garnished or soaked and cleansed before the stove or the gas oven is lit. Use of pressure cookers is another means of saving energy.

One need not dwell on the need to plant more trees; or the advantage to which kitchen waste can be used when one thinks of providing adequate manure for plants in the small patch at the rear of the house.

In other words, if every citizen, may be in a limited field of activity, makes use of the awareness of ecology and environment, this earth shall not turn into a death trap of pollutants.

Natural Calamities and Mitigating the Suffering

Developing countries are most vulnerable to disasters. High fatalities, meagre resources and fall in standard of living have caused economic setbacks. The natural way to tackle calamities is to integrate disaster reduction efforts and development planning at all levels.

Between 1900 and 1991, there were around 7000 events of different types of disasters, of which 40 per cent took place in the last three decades. From the 1960s to the 1980s there had been a five-fold increase in the frequency of disasters and six-fold increase in losses.

There is no reason why we should wait passively for the next natural disaster to strike. We should mobilise scientific knowledge and technological know-how.

1.66 billion people were affected the world over during the decade 1980-90 and of them 1.44 were in Asia. In South East Asia, India accounted for the largest number of events and of people affected although in terms of deaths, Bangladesh topped the list. Thus, India has a long history of coping with natural calamities.

The disasters most commonly faced by India are *drought, flood, cyclone, and earthquake.* The topographic conditions are such that one part of the country or the other is faced with any one of these disasters round the year. 68 per cent of the total sown area of about 140 million hectares are vulnerable to drought conditions and the drought management plan, evolved over the years, had proved quite effective.

The Indian drought management is based on an integrated strategy comprising *forecast and early warning systems*, contingency agricultural planning, effective accessibility to foodgrains for the affected population, protection of cattle and provision of drinking water and health care facilities. This has helped in minimising the crop loss and reducing starvation levels. For instance, during the 1987 drought, the crop loss was reduced to a mere 3.4 per cent over the previous year's output. In 1987 the starvation level was reduced 0.2 per cent from 36 per cent in the earlier drought periods. In the 1965 drought, we had to take care of 11.5 per cent of the population in the affected are whereas in the 1987-88 drought, only 2.5 per cent of the population required assistance. In 1965, 80 per cent of the funds were spent on providing relief whereas in 1987 only six per cent of the funds were spent on this account. The remaining 94 per cent were spent on development works.

A total of 40 million hectares of area are prone to floods and of them 13.5 million hectares have been afforded protection by means of *conventional structural measures like building of dams, embankments, drainage channels and river training works.* Of late, the emphasis has been shifted in favour of non-structural measures which include strengthening of flood forecasting network, timely warning and evacuation of the people to safer places. There are 157 flood forecasting stations covering most inter-State rivers. The timely dissemination of flood forecast to the administration at various levels has helped in the evacuation of people.

Besides drought and floods, cyclone is frequent. The 5700 kilometre coastline is exposed to hazards of tropical cyclones. On an average, two to three out of the six tropical cyclones in this region have been found to hit the Indian coast. Radars provide overlapping surveillance, thus ensuring that no cyclone in the vicinity of Indian coast can escape detection. Once it is tracked, warning is issued through INSAT satellite system and this helps in timely evacuation and consequent reduction in loss of life. The 1977 cyclone took a toll of 10,000 lives, whereas the 1990 cyclone resulted in just 910 deaths. The reason was that as many as six million people could be evacuated to safer places in time.

About 60 per cent of the total land area is vulnerable to seismic activities of varying intensity. These are mostly in the Himalayan region and Andaman and Nicobar Islands. Unlike in the case of drought, floods and cyclone, prediction has not yet been made possible in the case of earthquake. Hence the focus of mitigation efforts is on promoting *earthquake resistant structures.*

Modern tools and technologies are being harnessed for *disaster management. Space technology, for instance, has been used for surveillance of drought conditions and forecasting of floods and cyclones.* Many institutions are engaged in strengthening the programmes of training of health and other personnel, besides voluntary agencies, to deal with disaster situations in all the three phases — before, during and after the event. Education of the community assumes importance in this context.

The use of autonomous aerosonde is helpful to track cyclone movement. The aerosonde is conceived as a small aircraft with sophisticated avionics and other gadgets to take observations at the centre of the cyclone. This has been tested so far in Australia but it is claimed to be useful for tracking tropical cyclones over the Bay of Bengal. The design and development of the aircraft is expected to be completed by the end of next year. A major advantage of this is that the unmanned aircraft flies into the centre of the tropical cyclone and finds out much more about the cyclone intensity and the course it will take, than is possible by the techniques now being adopted.

While hazards are inevitable, disasters are not. The link between hazard and disaster can be broken by *risk assessment, prevention, preparedness, and emergency response*. This, in turn, will help minimise the economic losses the community suffers due to natural calamities. Currently, there has not been adequate evaluation of the economic impact of disasters, nor adequate awareness of the cost effectiveness of mitigation actions. The benefits of disaster reduction should be included into development projects e.g. disaster reduction features. Cost-benefit analysis of mitigation actions should be focussed in future. Such an analysis will help overcome the prevailing barriers to disaster reduction steps.

Developing countries are most vulnerable to disasters. There are high fatalities and declines in standards of living, through homelessness, health problems, malnutrition, loss of jobs, and displacement. Because resources are slim in developing countries, disasters cause burdensome economic setbacks. The best way to manage is to integrate disaster reduction efforts and development planning at all levels.

CONSERVING NATURE FOR THE FUTURE

Never before in the long history of planet earth have the ecological future and survival of life been as gravely threatened as they are today, thanks to the misadventures of humankind which created the technological revolution. Human beings are at the epi-center of this ecological destruction today.

Humankind has to ask itself what its destiny is going to be and how it can redirect its path to the future by developing plans and projects which can bring back the original harmonious relationship between man and Nature between environment and development, between present and future.

It is a hard process of transforming man's ways of living and methods of dealing with Nature. It involves redesigning the lifestyle, to bring back the sustainability which existed for millions of years.

Stark Realities: Unfortunately, the facts concerning sustainability are gruesome and tragic, threatening the very survival of humankind. In a poor country like India with its 890 million people and a land mass one-third the size of the U.S., the realities are as follows:

— Almost 50 per cent of the land is degraded and the situation continues to worsen irreversibly.

— Each year 6 billion tons of rich topsoil needed for agriculture is being lost.

— The food production of 185 million tons is not sufficient for feeding the population, thus driving at least 240 million people to the sub-hunger level.

— In large areas of the country subterranean water levels have declined almost five to ten times below the surface during the last 50 years.

— At least 35 per cent of the people get their cooking fuel from the firewood they collect by cutting down forest trees.

— Each year the country is losing, despite many afforestation programmes, at least 1.7 million hectares of life-sustaining forests.

— At the same time, the country has been adding 4 million more people than the total population of Australia to the land annually, that is 20 million people.

There are many more grim facts to be faced.

Fortunately, there are rays of hope too. Environmental education programme for people helps so that they would know and protect the environment.

Environment Act provides far-reaching legal framework for environmental protection.

There are efforts all over the country to develop agro-forestry projects, which address some of the fundamental ecological problems in an integrated manner. They provide opportunities for horticulture, water conservation, urban renewal, energy development, tree plantation, floriculture, soil regeneration and financial benefits.

These projects as well many other similar ones through their economic incentives and integrated ecological planning, would re-direct approaches to environmental sustainability.

Ecological Reconstruction : There are many such projects at the micro and macro levels, which provide some hope of ecological reconstruction. Obviously, the task is monumental, the resources limited the problems complex and the time limited. What is needed is a vision of renewed ecological integrity. Projects which address the problems at the grassroot level are needed. Programmes which are practical and feasible are the need of the hour — programmes in agro-forestry, in food production, in water conservation and in recycling waste.

We cannot simply think about development for the sake of development and profit only. Development which would benefit the poorest of the poor ought to be initiated. What is needed is planning and use of technologies which not only protect the ecological system, but which also enhance environmental quality.

The present environment crises are consequences of past follies — follies of greed, exploitation and domination. These have led to despair, deprivation, disease, disaster and probable doom. The present exigency demands that one acquires the basic wisdom of Nature, that is, to follow its laws and learn to manage according to its principles.

PRESERVING BIOLOGICAL DIVERSITY

Planet earth is endowed with a rich variety of life forms, and the teeming millions of these living organisms have been well-knit by the laws of nature. The inter-dependence of the various life forms starting from the uni-cellular primary producers to the complexly built higher plants and animals is a unique feature of this green planet.

Biodiversity as this assemblage of life forms is acknowledged as the foundation for sustainable livelihood and food security. Scientists have estimated that more than 50 million species of plants and animals including invertebrates and microorganisms occur on Earth, and hardly 2 million of them have been described by man so far. Scientists are also aware of the immense potentials of the various life forms.

Our biological future depends on our ability to conserve and utilise the rich genetic deversity occurring in living organisms on our planet. The importance of conserving the biological diversity; needs no emphasis.

Much of the spectacular successes in plant variety development of the rich industrialised countries are attributed to the richness of genetic diversity at the centres of origin and primary diversity of economic species, located in the poorer developing countries. While the genetic indebtedness of the north to the south is widely recognised, sharing the economic benefits accruing from genetic wealth is still a matter of debate and discussion.

The advent of the era of molecular biology and recombinant DNA research has brought home the point that all forms of genetic diversity have potential commercial value and therefore needs protection. The basic feedstock for the biotechnology industry is biodiversity.

The contributions of Nikolaivanovich Vavilov to the study of the centres of origin of cultivated plants will be a great interest to biologists and environmentalists.

HIMALAYAN ECOSYSTEM AND BIODIVERSITY

The Himalayas, are the youngest mountain systems, but have inspired mankind in all ages. As the marine fossil evidences from the

Himalayas shows, these mountain ranges might have arisen seismically out of the original 'Tethys Sea' that once separated the northern 'Angara and the southern Godwanah' land masses.

The Himalayan ecosystem and biodiversity are perhaps the most unique of their kind in the whole world and, therefore, saving them for our posterity is a global responsibility indeed. Unless their habitat is first of all protected and even enhanced, by not only India, but also by all the other countries bordering the Himalayas, the pride of these coutnries as well as that of humanity would soon be lost forever.

The profuse drainage system of the various mountain ranges of the Himalayas and the climate are the keys to the rich and varied habitat that supports one of the most magnificently diverse and populous vegetation and wildlife repositories in the whole world. About 117 species of mammals, 983 species of birds, 122 species of reptiles and 20 out of the nearly 200 species of fish occur in the Himalayas. There is an infinite variety of medicinal herbs, shrubs and fruit trees, lower animals like worms, insects and spiders that constitute the basic food-chains for the rich wildlife.

In order to conserve this rare and valuable wildlife of the Himalayas, 19 national parks, 58 sanctuaries, three biosphere reserves and four reserve forests, some of them being the world famous one, are maintained whose geomorphology, climate, forests and wildlife are well-studied and accounted for.

The Himalayan ecosystem and its wildlife are precious and their conservation programmes need sustainable development. (P.J. Sanjeeva Raj)

SUSTAINABLE INDUSTRIAL GROWTH AND POLLUTION CONTROL

Industry, is central to all economies of all societies and an indispensable motor of growth. It is all the more essential for a developing country to widen its developmental base to meet the challenges of exponential growth of its population. The essential needs of vast multitude of people have not only to be met but also legitimate aspirations for improved standards of living have to be created on sustained basis since the satisfaction of human needs and aspirations is the primary objective of development. Sustainable development is development which meets the needs of the people *without unduly disturbing the ecology.*

In order to *promote sustainability of development* which is intimately linked to the dynamics of population and growth, we must define the *environmental goals and development objectives* and establish clear priorities among competing demands on their scarce resources. When industrialisation is accelerated, then only it would be possible to create a wider service sector which provides a major employment potential to our educated youth. Unemployment is one of the persistent unmanageable problems facing us today. Therefore it should be the endeavour of our planners to remove unemployment and solve under-employment in a time frame. All this calls for a co-ordinated efforts between industry and state to work together, then only economic development and growth can be achieved.

We are all aware that a country like Japan, which is deficient in natural resources, achieved economic miracles inspite of limited land, water and other resources, thro' integrated economic planning and collective effort between the Government and industry. But the existence of abundant resources itself is not enough without advance and proper planning for exploitation by modern technology and required infrastructure facilities so as to utilise the resources to induce maximum input/output ratios. Human progress has always depended on our technical ingenuity and a capacity for co-operative action.

Economic development is a complex process. To create a sustainable industrial growth, several factors have to be taken into consideration by harmonising them in such a way that it leads not only to economic development of the people but also helps in preserving ecology and environment. Environment and developemnt are not separate challenges and they are inexorably linked. All these call for an integrated working beween the industry and Government.

Pollution abatement is a 'Techno-Economic-Complex' problem and solutions not easily discernable. While realising the need for preserving and maintaining our ecology and environment, it is necessary to realise the limitations and the restrictions in finding solutions to this complex problem. Strict pollution control does not necessarily mean total pollution control which is not possible. Even in countries like Sweden, Finland, Holland etc. total pollution control is not attempted, for it involves employment of huge capital resources which no country could afford. However, since the ecology and environment have to be preserved, our effort should be in the direction of minimising the problem to the extent possible within permissible economic limitations. Economically the cost of pollution control soars as pollution standards become more severe. Therefore, every country has to evolve its own pollution standards, depending upon its own land resource as well as the level of industrialisation, nature of industries and the receiving water system, etc. Finland, it is said, a land of lakes, having 55,000 lakes, set up large industries like Paper, chemicals etc. without causing hardship to the society by solving the problem of the pollution abatement and discharge by honest interaction, co-operation and co-ordinated effort between the Government and other agencies concerned.

As a First Step twoards this objective, the Pollution Control Board and other Government environment authorities should stipulate reasonable and attainable standards taking into consideration the over all impact on environment. Hence, a time has come to plan and propose a long term environmental strategies for achieving sustainable development by the year 2000 and beyond and help various Government authorities to define shared perceptions of long term environmental issues and the appropriate efforts needed to deal successfully with the problems of protecting and enhancing environment at the same time allowing increased industrialisation. The *two problems viz. widening base of industrialisation and ecology preservation cannot be viewed in isolation.* There is a strong need for integrated approach and they are not separable. The need for prescribing a phased achievement of standards is not something new.

The standards attainable by the pollution technology have to be cost effective in terms of public health and industrial economy. The technologies should offer the promise of higher productivity, increased efficiency and decreased pollution.

The possibilities of solutions can be many depending upon the choice the society makes (e.g., as regards the uses of waste waters). Every development brings along with it certain problems for which solutions have to be explored. At times a compromise may be necessary in favour of industrial growth if overall benefits to the society arising out of it are more than those obtainable with higher quality of streams/ drains. The availability of technology within the country at permissible costs is an important consideration. Thus, a number of factors are required to be considered while making decisions in matters related to pollution control and abatement. *Extensive use of chemical fertilisers, pesticides, insecticides for growing more food impairs the quality of our waters in our rural settings to-day but still we use them in the larger interests of food production for our growing population.* Today's needs tend to overshadow consideration of the environmental future. So the case may be of industry and its concomitant pollution problems.

Industry should accept the legal responsibility and social obligation of the industry to treat the effluent generated from units within the available technology and economic feasibility in the best possible manner. But after completing the treatment plants, the major problem that comes before us is regarding discharge of effluent into streams/drains. After treating the effluents to be a tolerable level, it is to be discharged into streams/drains that may be designated and declared by the Government.

With the rains carrying backwater from the fields to sea coast, it should be possible to effectively manage the treated waste water from the industries and carry it to the coast and ultimately leaving into sea for absorption. Instead of allowing industries to discharge the treated effluent water in every drain of their choice, the Government itself should identify after careful study and earmark one or two exclusive drains in each district for carrying the treated effluent water by different industries. The discharged water into drains normally is highly diluted and it is fit for agriculture purpose and is proved since many years and it is worthwhile to contemplate to use the diluted water in the drains for utilisation of agriculture where adequate irrigation facilities are not available. This practice of letting out of effluent water after treating the same into waste water drains is adopted even in developed countries.

In our own country, where major industries like Paper, Chemicals, Fertilizers, Pharmaceuticals, etc. are set up, the discharge of industrial effluents is made by exclusive drains carrying discharged waste water. Thus, by providing exclusive drains for carrying treated industrial waste water in rural areas of our districts, it would help us augmenting economic development by setting new industries in the area based on our resources available and also now potentially promising gas and oil. What is needed now is new era of economic growth — *growth that is forceful and at the same time socially and environmentally compatible.*

ECOLOGY AND SUSTAINABLE DEVELOPMENT

Awareness of ecology has three main dimensions (a) the rapid degradation of environment, (b) the enormity of the problem and (c) the challenges in finding a meaningful solution. The world is still to strike a balance between industrial and economic development and ensuring a livable environment to the people. Unless effective steps are taken in time we may not have pure water to drink, fresh air to breathe and non-contaminated food. Governmental and non-governmental action to stem the rot has not been keeping pace with the rapidity with which the habitable surroundings are degrading. Each section of the population is contributing its share to this environmental assault. If the rich countries have unbridled lifestyle at the cost of LDCs, the ecological problem in the poor countries is compounded by the exponential population growth and ill-balanced economic development.

The agents of pollution are: depleting ozone layer, global layer, global warming, use of pesticides, heavy metal toxicity, industrial effluents, smoke emanating from factory chimneys polluting etc. As regards aquatic environment, it has been said that there is an urgent need to restrict discharges of effluents in the water courses and to reduce the loading of pollutants and increase the self-purifying capacity of the water network. How several industries like mining, cement, paper mills and the like have been responsible in polluting air and water is well-known. The lurking dangers of pollution by mercury through discharge of industrial wastes arises from the fact the mercury so discharged may remain in the environment. These deficiencies tell on the health of the people. Several persistent diseases occur among the rural and tribal population.

The relationship between man and environment appears as a paradox. Nature simultaneously appears to be sustaining and unyielding. Does man still have an edge over nature?

SUSTAINABLE DEVELOPMENT

Scientific methods should be evolved for using natural resources sparingly for sustainable development.

Equitable use of common resources requires much greater understanding of the complex forces which create local as well as global environmental problems. *These forces derive from population growth, particularly in developing countries, over-consumption in affluent countries, misuse of natural resources, land degradation, deforestation, loss of biological diversity, rapid industrialisation, waste accumulation, excessive water and energy consumption and possible climate changes due to anthropogenic impacts.*

The earth is simultaneously suffering from underdevelopment and overdevelopment and called for redistribution of military spending towards planetary development, projects, many of which would involve the transfer of technology from developed countries to developing countries, and the introduction of the ''Polluter pays principle''.

''The issue is not merely one of transferring know-how from one country to another, or one enterprise to another. It is a question of building up capacities, both technological and institutional. It is a question of ensuring co-operative research on science and technology.''

TRAGEDY OF DEFORESTATION

Hazards of Forest Clearance

Once the land is cleared and replanted, it is not the original forest that grows again and may well destroy the original gene pool. It is difficult to reclaim for land forest again. (Gradwohl and Greenberg)

Clearing and burning is likely to release most of the nutrients into poor tropical soils which are then leached out by warm tropical ruins.

The process of forest regeneration is complex and seedlings of hard woods found in the forests often tolerate only a narrow range of humidity and light conditions and cannot therefore, grow in open areas. Under natural conditions a succession of different species, starting with the coloniser, provides this environment.

Seeds are most often spread by animals sometimes by relatively few species. This means that the seed source has to be near the clearing. And that the dispenser populations must be able to withstand drastic habitat changes.

Forest trees are often pollinated by animals breeding systems such as dioecy (where male and female flowers are on different plants). These plants require cross pollination between individuals and this means that they must grow in proximity with the proper pollinators in the area.

Both pollination and dispersal are made more difficult by relative rarity of tropical forest species. This rarity is a natural consequence of diversity, since tropical forests unlike many temperate ones are not usually dominated by a small number of species.

Trees may require specific symbiotic relationships with fungi known as *mycorrhizae*.

There is a collection of ungerminated seeds in tropical soil but it is unclear how long the seeds remain viable. The degree to which seeds can remain viable in tropical soils in the face of insect and fungal attack varies greatly between species. *Further the conditions found in large human generated clearings are often too stressful for soils to survive.*

Tough, fire resistant grasses and shrubs take hold after long periods of burning and grazing. These plants can effectively block forest regrowth.

Even when forests are managed it is found that the original composition of the forest is not maintained giving rise complications.

Hazards from Hydro Electric Dams

Hydro electric dams, in the name of development destroy thousands of square kilometres of tropical forest. The process of filling the reservoirs may drown large tracts of forest, displace people and kill wildlife. Although dams are intended to provide inexpensive electricity, *many are economic failures because of lack of environmental planning.* Erosion of water shed fills reservoirs with silt and reduces the ultimate output and useful life of dams.

Deforestation

The consequences of deforestation according to Gradwohl and Greenberg can be summarised as :

— Loss of people and cultures whose way of life depends upon the forest along with the loss of their knowledge;

— An increase in barrenland and desertification in drier tropical regions;

— Regional decreases in rainfall, excerbating desertification;

— *Global increases in temperature due to a rise in atmospheric carbon leading to a rise in sea levels.*

Forest vegetation protects soil from the full force of tropical rain storms. The canopy receives the brunt of rain drops that are further broken up as they hit the undergrowth but if the forest is cleared especially when the vegetative cover on the soil surface is removed, the bare earth receives the full force of intense cloudbursts and the top soil is washed into the nearby streams. This loss of soil causes two distinct problems :

— It is difficult to reestablish any sort of vegetative cover either in the form of crops or trees;

— Displaced soil which enters the river disrupts steam ecology and increases loads in dams down stream. (Parimala Appasam)

USE OF COMPUTERS FOR MIS IN FOREST CONSERVATION

The advantages of the computerised Management Information System (MIS) in forest conservation is multifold. The time saved by staff in clerical work is drastically reduced. It liberates the field staff from office work so that they can spend more time in the field. Repetive recording of data is eliminated. Data retrieval efficiency is substantially improved. It would facilitate the timely submission of statutory reports. These reports need not wait till the month-end but on any specific day.

Once the basic documents are input, the preparation of reports can be done in minutes. Another advantage of this system is the variety of reports generated. These can be generated in any given format required. The information is processed so that a progress report can be generated scheme-wise, range-wise or mandal-wise.

The system is an effective monitoring tool. At any given moment, the physical or financial progress reports can be printed. For circle meetings, district or state level meetings, appropriate reports can be generated at short notice. The system is thus a time saver to operational level staff and a monitoring and control tool to the management.

A Management Information System has 3 major pre-requisites the hardware, the software and the human resource. All three should be adaptable in the existing environment.

Hardware involves the selection of appropriate technology. In the forestry sector, the hardware has not been a problem.

Computers have invaded most offices. But it has remained merely as a cosmetic accessory, used at the most for word processing. The key to the conversion of this hardware into the powerful information processor that it is, lies in designing appropriate software.

Appropriateness of the software can be gauged in this context by three major parameters:

— Acceptance by the immediate user; whether it fits into the existing operational system; and policy level acceptance.

— The second and third parameters are essentially technical challenges. While introducing such an MIS at the grassroot level, the needs of the grassroot level user become imperative. There could be an initial diffidence towards computerisation. Mythical fears of the complexity, rigidity and manoeuvrability of the computer were identified as the main causes of rejection. The challenge was to design a system without these shortcomings and to communicate this effectively to the ultimate users.

Simplicity, flexibility and comprehensiveness are the three hallmarks. Simplicity essentially means user-friendly. The model is so explicit in this context that it requires no prior orientation of the user to computers.

It eliminates the necessity for keying in data. All operations are done merely by selections from displayed data/functions.

Flexibility is the next hallmark that motivates a user to use the package. The basic fear of computers is the restrictions or boundaries that it may set to the multifarious human activities. This trap has been avoided with a modular design of the system which facilitates flexible operatability. Extension is as much as important aspect. The user needs to be convinced that the computer shall place no restrictions on his activities. It is flexible enough to adapt to his varied requirements and would only help hasten the process. The fact that with knowledge and skill, the operator would be the master over the computer is specifically emphasised. The software is so designed that any number of new schemes or project components can be introduced very easily.

Comprehensiveness is a mammoth task considering the enormous piles of paperwork in the department. However, this package deals with the problem from the root. The package concentrates on the three basic documents of voucher, sanction order and estimate. The entire accounting system of the forest department rests on these basic pillars.

Voucher		Statutory Reports
Sanction order	MIS	MIS Reports
Estimate		Control Reports

The entire gamut of reports required by the department is generated using these three basic documents. No additional forms or data are needed to be keyed in for the system to function.

The third stage is the development of the human resource. Individuals working the present system have to be convinced, motivated and trained to utilise the new system.

Development of the human resource is a major challenge. People who have grown with the existing system show a natural reluctance to accept a change. Computerisation is feared to destabilise them. The first step in training is to defreeze this attitude. Diffidence has to be overcome with appropriate information dissemination. The system should be designed in such a manner that existing staff at the divisions can operate the system.

The learning of skills begins when this defreezing has occurred. The challenge for the future lies in refreezing this change as a way of life. For this monitoring and periodic refresher courses are a must. Teething problems are bound to arise. Tackling these should be top priority. (Varadan)

HOW TO REVIVE NATURAL STREAMS FOR SELF-CLEANING AND WASTE ABSORPTION

Throughout the world, small streams and rivers have for years been the object of "improvement" efforts. Such projects unquestionably increased the amount of land available for farming. But, it is now realised, they also destroyed many functions which natural streams and streamside vegetation perform, that benefit mankind-functions we can ill-afford to neglect.

One of these is the ability of streams to absorb urban, industrial and farm waste, something they have been doing for centuries. Not only do streams carry waste away but, if they are not overloaded and their physical structure is not destroyed, they gradually reduce the concentration of nutrients, organic compounds, metals, and toxic pollutants such as pesticides, herbicides and excess chemical fertilizers that run off from farm fields. In some places where the ability of streams to do this has been compromised through channelisation, too much nitrate is leaving farms and entering the groundwater, where the resulting high nitrate levels pose a health hazard to drinking water.

In addition to this so-called "self-cleaning" ability, a healthy stream provides habitat for fish living in the channel, as well as wildlife and medicinal plants living along its margin. In arid and semi-arid areas, the shading canopy of streamside vegetation can effectively reduce water temperatures, thus minimising moisture loss through evaporation.

An adequate solution to the resulting problems can be implemented at the landscape level, involving entire catchment areas, and including, not only the channels themselves, but streamside vegetation zones as well. A complete change in the structure of the landscape is needed, from a patchwork of fields to a mixture of fields and clean-flowing streams protected by a streamside canopy of vegetation. This necessitates a plan that can take into consideration vast differences in local topography, economic resources, and personal preferences of individual landowners. As far as possible, such a plan must also be simple, cheap and flexible.

Cost/Benefits of Small Stream Restoration

Restoration measure	Cost	Benefits
1. Buffer strips	Purchase or rent of two hectares of land per km of stream	Base for restoration Plant habitat Stream bank
2. Revegetation	None for natural colonisation Cost of plants	Faster protection afforded Channel shading
3. Slope reduction	Removal of $2m^3$ of soil for 50% slope reduction	Reduce erosion Water quality protection
4. Meanders	Digging of meander channel	Increase stream length
5. Riffle-pools	Stones to place in stream	Improve re-aeration fish habitat
6. Ponds	Construction of earth dam	Water retention habitat Aquaculture Wildlife
7. Wetlands swamp forests	Cost of land to set aside	Water retention Wild life habitat Improve water quality

One such approach is to see restoration measures as a series of building blocks (see table), which can be combined in different ways over time according to landowner preference, landscape form and available funds. Installed together, these blocks can improve natural habitat, restore self-cleaning capacity, and reduce the loss of important nutrients from the landscape. While some measures are more important or costly than others, all are effective. They can be used individually or together.

Technology

The technology required for stream restoration is small-scale and has low maintenance and operating costs. This makes it especially useful for farmers in developing, as well as developed countries. It is based on commonsense notions, which anyone living close to the land and depending on its long-term quality can readily comprehend.

The orientation of the common measures are aimed primarily at lowland streams, but the approach can be used to restore any stream and should be seen as a general strategy for both landowners and governments. Other restoration measures specific to regional problems and topography can be added to those already existing.

STUDY OF BIOSPHERE AND GLOBAL CHANGES IN WEATHER

The international Geosphere Biosphere Programme, will gain an insight into how land, oceans, and the atmosphere are affected by the global changes in weather. It will strive for improved scientific knowledge of the sources, causes and impacts of climate change at the global, regional and local levels. A wealth of information now exists to show how the human species is influencing the life-support systems of planet Earth, "Such human impact will inevitably intensify in the next few decades, with a strong likelihood that they will cause disruptive changes in climate and other global functions". The objective of the IGBP is to reduce the uncertainties inherent in predicting the future consequences of current trends.

The tremendous improvement in computing power in the last decade has resulted in designing circulation models which provide an insight into how the world weather system works. A global, three dimensional simulation of the dynamic behaviour of the atmosphere, after testing against observed events and past data, points to some disturbing features. If the current trends of increasing emissions of greenhouse gases continue, average global temperature will increase by 1.5 to 4.5 degree celsius in the next century with an accelerated rise in global sea level, warn the experts.

Global warming inevitably brings about alterations in rainfall patterns and influences the earth's water cycle. On an average, water resides in the atmosphere for about ten days before returning to the earth's surface as rain or snow. What is really fascinating about the IGBP is the integration of traditional scientific subject areas on an unprecedented scale bringing together researchers in biology, physics, chemistry, geology, and other disciplines. The use of state of the art technology for high *precision analysis, network computing and satellite remote sensing has made* the IGBP a remarkable international venture.

Adverse Effects of Climate Changes

Gobal climate should be assessed based on the implications of the changed *weather conditions on agriculture and forestry, natural terrestrial ecosystems, hydrology and water resources, ocean and coastal zones and terrestrial cryosphere.*

II

ENVRIONMENTAL DEGRADATION AND ITS DESTRUCTIVE EFFECTS

Ozone Depletion can Reduce Plant Photosynthesis

Hitherto, we have been hearing of higher incidence of skin cancer, cataracts and problems with the immune system in humans due to the increased amount of Ultraviolet-B (UV-B) radiation reaching the earth through the depleted ozone layer in the stratosphere.

Now comes the disconcerting news that the increased UV-B radiation can reduce the effectiveness of plant photosynthesis as well as multiplication of phytoplankton (the plant organisms at the bottom of the food chain in the oceans).

In laboratory experiments, an Australian research team took measurements of six performance indicators for the photosynthetic process using ordinary garden peas (*Pisum sativum*) after they had been exposed to UV-B. The result: There was a reduction in performance of all six with some declines as much as 70 per cent.

The research was prompted by the discovery of an area of ozone depletion (the infamous hole in the ozone layer caused by chlorofluorocarbons used in aerosol sprays) over Antarctica. The depletion is a seasonal phenomenon in the southern hemisphere occurring during a crucial period for plant growth — spring and early summer. It allows increased levels of damaging UV-B light to reach the earth's surface in the high latitudes of the southern hemisphere, where temperate food crops are grown.

Antarctic biologists are showing that there are effects on plant plankton in the southern oceans due to increased exposure to UV-B and this has major implications for the productivity of the fisheries in the region. Fortunately, nature has an antidote to combat this adverse effect of UV-B on photosynthesis. Some plants produce, in the superficial tissues, a class of substances called Isoflavonoids which can act as a UV filter or screen. Some plants do this better than others. Research has now to focus on how these plant species produce this sunscreen so that plant breeders can develop UV tolerant strains.

"Rain forests and other groups of tropical plants are likely to contain some of the answers that we require because they already cope with higher ambient levels of UV-B".

OZONE HOLE OPENS EARLY, CLOSES LATE

Springtime heralds many things among the more depressing being the appearance of a 'hole' in the ozone layer above the Antarctic. New findings may turn mild depression into fathomless gloom, because it now seems as though ozone destruction starts sooner and lasts longer.

Ozone depletion is caused, in the main by a substance called 'chlorine monoxide'. This is brewed in the upper atmosphere from manmade ingredients, the chlorofluorocarbons (CFCs). But there's a catch: chlorine monoxide can only exist in the catch: chlorine monoxide can only exist in the presence of the so-called pollar stratospheric clouds, which form at extremely low temperatures. Without the clouds, chlorine remains bound up in inactive 'reservoir' molecules which cannot destroy ozone. This explains why ozone depletion tends to happen in the polar regions rather than elsewhere. A large amount of chlorine monoxide was evident in the Antarctic atmosphere, which should have caused a sizable loss of ozone then and there.

It could be that chlorine monoxide is not the prime culprit for ozone depletion after all. If so, why do all other results implicate chlorine monoxide so strongly? Alternatively, it could be that ozone was indeed depleted by chlorine monoxide, but that the effects were masked by the influx of ozone-rich air from higher altitudes during the winter. If this explanation is correct, last year's ozone loss over Antarctica must have started sooner, and lasted longer than anyone has realised.

Because the Antarctic is much colder than the Arctic, stratospheric clouds can exist there for several months during the southern winter and spring, and the ozone depletion is severe as a consequence. Fortunately, higher temperatures in the Arctic mean that polar stratospheric clouds are less common, and as yet there has been no significant springtime hole over the Arctic.

The aim was to continuously monitor the important chemicals that take part in the complex interactions that deplete the ozone layer. Before this, many of these chemicals have to be measured locally from balloons or aircrafts, and it was difficult to fit these individual pieces of information together to form a coherent picture. The satellite can only focus on one hemisphere at a time, but every 36 days or so it switches its gaze, enabling it to follow the formation of the Antarctic ozone hole and to see whether there is any chance of a similar hole forming in the northern hemisphere.

The instrument used is called the 'microwave limb sounder', and it measures ozone, chlorine monoxide, pressure and temperature in an attempt to pin down the events leading to ozone loss.

GLOBAL WARMING THREAT

The picturesque coral reef islands face threat from rising sea level triggered by the greenhouse effect.

The islands are low lying with the highest point on the islands not more than a few metres above sea level. A rise of one or two metres in the mean sea level can submerge the islands. Study of the rates of global warming due to greenhouse gases shows that it has triggered an increase of 10-20 cms rise in sea levels. The estimates for 2050 AD vary between 23.8 and 116.7 cms.

Global warming can arm the cyclones with more destructive power and higher frequency of occurrence. According to a study if the amount of carbon dioxide in the atmosphere doubled form the present levels, the destructive power of cyclones will be 40-50 per cent more.

IMPLICATIONS OF CARBON LIMITS

To determine the international effects of carbon constraints, Alan Manne and Richard Richels have analysed the proposal to stabilise global carbon emissions by imposing a 20 per cent cut-back in industrialised nations while allowing developing countries to increase emissions by 50 per cent.

Even with this differential to allow for faster economic growth, developing countries would eventually pay a higher price in terms of reduced gross domestic product (GDP).

For the U.S. and other industrialised nations, the cost of emission controls would run between one and three per cent of GDP for most of the century. The costs to China and other developing countries would rise to 5-6 per cent.

"The costs are higher for developing countries because their economies are growing faster," explains Richels. "Both population growth and economic expansion will continue to put upward pressure on the energy demands in these countries. And the cost of any restriction of carbon emissions will fall most heavily on countries, such as China, that are hoping to fuel economic expansion with an energy resource base dominated by coal."

Global patterns of CO_2 emission will also shift significantly under the assumptions of this analysis. In 1990, the industrialised nations accounted for 64 per cent of total emissions; their contribution is expected drop to 30 per cent by 2100. Since economic growth rates are considerably higher in the developing countries, those countries will also have higher growth rates in energy consumption and carbon emissions.

PASSIVE SMOKING CAN KILL

The environmental protection Agency (EPA) has found that as many as 3,000 non-smokers may die a year from lung cancer because they inhale the smoke of people they live or work with.

Smoking is not just health danger for smokers, but a significant risk for non-smokers, particularly children who are exposed to secondhand smoke.

Scientists studied 30 reports from around the world in coming to the conclusion that secondhand tobacco smoke produces cancer.

The EPA reports state that about 434,000 persons die each year in the United States from cancer, heat diseases and respiratory ailments caused by smoking. "Smoking is the leading preventable cause of death today."

Both the smoke rising from cigarettes and that exhaled by smokers contains a complex mix of more than 4,000 compounds, a mix that "contains many known or suspected human carcinogens and toxic agents".

Passive smoking particularly hard on children, contributing to 150,000 to 300,000 cases annually of respiratory illnesses such as pneumonia, bronchitis and bronchiolitis among children less than 18 months old. That results in 7,500 to 15,000 hospitalisations a year. Passive smoking worsens the condition of 200,000 to 1 million asthmatic children.

There are more than 10 times as many heart-disease as lung-cancer deaths caused by passive smoking.

We have far better data and more persuasive data on lung-cancer.

In Eastern Europe, however, the current proportion of smokers is 70 to 80 per cent. Smaller proportions smoke in developing countries but the rate is rising.

"China has a great problem because its population is large and smoking may lead to millions of deaths a year."

Only a few developed countries — the United States, United Kingdom and Australia — have effective anti-smoking programmes. Even fewer have begun to worry about the effects of passive smoking.

AIRBUS FLEET TO STUDY GREENHOUSE EFFECT

Scientists measure ozone and water-vapour at airliner cruising heights to gain more knolwedge of the "greenhouse effect". The measurements will be the first and most widespread of their kind in this part of the atmosphere.

Parameters

The project involves a fleet of five specially-instrumented four engined A340 Airbuses taking samples over a wide range of northern and southern hemisphere air routes.

Each of the giant airliners will carry about 80 kg of instruments that will automatically record ozone and water vapour levels every four seconds — equal to every kilometre or so at cruising speed — as well as position, pressure and wind.

Flights will begin in the third quarter of this year and measurements will continue to be taken in order to cover seasonal variations. The results will be analysed and made public a few months later.

These trials are the idea of Airbus Industries building the family of twin and four engined Airbuses. The project is the result of concern by scientists that the *Earth's atmosphere may be gradually warming up because of increasing concentrations of gases such as ozone and water-vapour at airliner cruising heights. This could lead to more of the Sun's energy being trapped in the atmosphere by the 'greenhouse effect', and have a major effect on world climate.*

EC Project

The results should help to clarify the debate about what is actually *happening and, through inclusion in atmospheric models, lead to more accurate predictions about our future climate.*

POLLUTION FROM ELECTROMAGNETIC INTERFERENCE — THE DANGEROUS WARNING

A silent threat is one the prowl. Emitted by various electronic and electrical equipments, it is pollution caused by the electromagnetic interference (EMI). By and large ignored for many years, its effect was ironically, felt way back during the Second World War. However it is now acquiring importance due to man's increasing dependence on these equipments.

Most electrical and electronic equipments including microwave antenna towers, switches, high voltage lines, radars, transceivers and electric motors emit electromagnetic radiation either intentionally or unintentionally.

The most common examples *at houses are the chokes fitted to fluorescent tube lights and mixer grinders. These emit radio waves which interfere with the radio sets resulting in 'boating'.*

Sometimes a scooter passing by disturbs the television reception. *It is because the electromagnetic (radio frequency) have emission, called ignition noise, from the spark-plug is picked up by the television antenna along with the actual signals.* This causes interference in the video. One ignores these disturbances as they are short-lived and do not have any serious consequences. These interferences are classified as unintentional since the emission of electromagnetic field is not a part of their function. Walkie-talkies and the like relay signals intentionally to serve a purpose. However *these radio waves can also interfere with, say telephone lines. Atleast those who reside near radio stations experience its interference and have to add special electronic filters to their telephone receivers to get a clear audio tone.*

EMI is of two kinds — conductive and radiative. In the conductive type their interference mostly occurs via the power lines as spurious voltages whereas in radiative type electromagnetic waves are set up in its environment.

In modern offices, hospitals and industries many hightech equipment are put together in a room and have become an indispensable part of the everyday life. Each of these gadgets sets up a field around it. These fields can interfere among themselves and may even disable other equipment. In industries, this crisscrossing may be catastrophic if the equipments are not compatible. Some reports say that EMI can make computers hang — without responding to the user — and may have to be restarted resulting in loss of man-hours. *In hospitals, this may lead the doctor to treat his patient differently.*

EMI field have no proven direct ill-effect on humans, though this cannot be ignored. Statistical reports say residents near electrical utilities tend to be psychologically affected and become tense easily. According to another report even the fields set up by computer screens increase the body temperature (by two degrees celsius) and make the heart beat faster to cool down the body. Consequently, *the eyes or genitals age faster and in severe cases lead to cataract or impotency.* Researchers have concluded that the *effect on humans depends on the intensity (voltage) of EMI field, its frequency and hours of exposure.* At high frequencies even low voltages are hazardous whereas at low frequencies only high voltages are serious.

Microwave Range

Another major source of pollution are microwave towers which have proved to have significant effect on humans. Some city development authorities have already made it a rule not to allow high rise buildings within the 'view' of these towers.

The use of microwave ovens in the kitchen is increasing. These emit fields at high frequencies, in *gigahertz range.* Reportedly these emissions coincide with the absorption frequency of human cells and, therefore, *sufficiently high exposures may even kill a human foetus in the womb.* Therefore, *pregnant women should keep away from these ovens.*

Satisfactory performance of electronic gadgets calls for certain design considerations. For example, any closed loop in a printed circuit board (PCB) will start emitting EMI fields and hence has to be avoided. CEPTED has the imported computer software and design facility to make 'good' designs using computer simulation. So the manufacturer can try his design, test it for EMI/EMC, modify it if necessary and arrive at a clean gadget. He can directly go to the production stage from the drawing board, saving time and money.

For safe operation various subassemblies within the equipment must be adequately shielded and packaged so that their concerted interference is minimised. *The components and subsystems also should be assembled so that there is no mutual transfer of heat affecting their performance. Thermal design at the PCB level is important.* (P. Sekar)

EMI

Configuration	Cause and effect
Space : Rocket launching sites have radars and other communication devices	These equipment emit powerful electromagnetic waves which are picked up by the rocket which may emit the nose cone prematurely.
Aviation : Aircraft is fitted with an array of microprocessors and radars.	A passenger switches on a FM radio. Its local oscillator couples to the radio compass equipment making it erroneous.
Marine : Aircraft carriers have powerful communication antenna and missiles on its deck.	The missile's electronic circuits suffer interference from the antenna and the missile tends to go off-course.
Police: Wireless network	Automobile ignition causes white noise which interferes with the mobile receives at busy road.
Industry: Microprocessors are used in process control	Fluctuating electrical loads induce voltage surges in the power line. The microprocessor assumes these as warning signals and may stop the process.
Hospitals : High tech equipment such as heart beat monitor are placed along with other equipment.	Operation of lifts, heaters or the like causes voltage fluctuations beyond the tolerance of these equipment making them go-off-beat.
Kitchen : Microwave oven	Radiates microwaves. Heart patients fitted with pace makes suffer discomfort.
Business houses: Cash registers are connected to the same power source which operates heavy motors.	Power source is contaminated by the operation of lifts, air conditioners, neòn signs etc. The cash register memory is erased.
Television : High rise antenna is connected to the TV by long leads	The video suffers from several wave patterns from vehicles, chokes, motors, welding shops and cable TV network.

FIBRES TO FIND METHANE IN COAL MINES

Scientists have produced an effective means of detecting the dangerous methane gas in coalmines. Researchers have produced a new optical fibre detector that promises improved reliability at lower cost than existing techniques that are said to be based on unreliable electronic systems or expensive and cumbersome optical systems.

Conventional fibres also allow the optical source and all electronic components to be located hundreds of metres from the monitoring area, thus providing safe and accurate monitoring of previously inaccessible areas.

The build-up of methane gas in mines is a continuing hazard because it can form an explosive mixture in air. The gas is odourless and colourless, and as miners feel no shortness of breath when concentrations reach critical levels, it is impossible to tell when methane is present without using some sort of detection device.

STRATEGIES FOR RESCUING THE GLOBAL ENVIRONMENT — LAWS, SYSTEMS AND PROCEDURES

Al Gore has charged a five-fold course of action to save the environment — a course that should be organised around five strategic goals which are:

- Stabilising world population
- Creating rapidly and developing environmentally appropriate technologies
- Changing comprehensively and ubiquitously 'economic rules of the road' by which the impact of decisions on environment are measured.
- Negotiating and approving a new generation of international and environment-related agreements that must be especially sensitive to vast differences of capability and need between the developed and undeveloped nations.
- Establishing a cooperative plan for educating the world citizens about global environment.

It will be appropriate to restate here the itemised economic rules of the road that Al Gore has listed out:

1. The definition of GNP should be changed to include environmental costs and benefits.

2. The definition of productivity should be changed to reflect calculations of environmental improvement or decline.

3. Governments should agree to eliminate the use of inappropriate discount rates and adopt better ways to quantify the effects of our decisions on future generations.

4. Governments should eliminate public expenditures that subsidise and encourage environmentally destructive activities.

5. Governments should improve the amount and accuracy of information on the environmental impacts of products and provide it to consumers.

6. Governments should adopt measures to encourage full disclosure of companies' responsibility for environmental impacts of products and provide it to consumers.

7. Governments should adopt programmes to assist companies in the study of the costs and benefits of environmental efficiency.

8. Nations should revise their antitrust laws to encompass environmental harm.

9. Governments should require the incorporation of standards to protect the environment in treaties and international agreements, including trade agreements.

10. Environmental concerns should be integrated into the criteria used by international finance institutions for the evaluations of all proposed grants of development funds.

11. Governments should make accelerated use of debt-for-nature swaps to encourage environmental stewardship in return for debt relief.

12. Governments should develop an international treaty establishing limits on CO missions by country and a market for the trading of emission rights among countries that need more and countries that have an excess amount.

ENVIRONMENTAL LAWS

Perhaps more fundamental as barriers to the operation of laws to protect the environment, than what may be seen as "Procedural" problems, are certain concepts which in fact illustrate the complexity of the environmental issue. These are "state sovereignty," "permanent sovereignty over natural resources," and what developing states claim as their "right to develop". All states espouse the first two concepts, but even here, the developing states are more vehement, doubtless because they have (in most cases) acquired sovereignty of any sort so recently.

The concept of state sovereignty implies a certain independence of action for all practical purposes. It means that within its territorial limits, the state is free to do what it pleases. State sovereignty is firmly entrenched in international law, and yet it creates problems in the environmental field. It means that the international community theoretically has no right to question activities within a state, yet such activities, in particular the degradation of the domestic environment, may create problems for the international community — e.g. the creation of a refugee problem because of environmental degradation.

Further, with the rise of global environmental problems, it is increasingly the case that purely domestic activities such as the utilisation of chlorofluorocarbons (CFCs), impact negatively on the global environment (in this case the ozone layer). While the international community would like to curb such activities, it is constrained by the principle of state sovereignty (and the allied notion of 'non-intervention' in the internal affairs of states). Conference on the Human Environment, which declares that states have 'the responsibility to ensure that activities within their jurisdiction or control do not cause damage to the environment of other States or of areas beyond the limits of national jurisdiction, was in fact mitigated by the opening words of the same principle.

The 'right to develop' is another potential and often real obstacle to environmental protection in the developing countries. Notions of 'sustainable development' are not, as yet, sophisticated enough to prevent a country claiming that logging is essential to its economic development. Despite its apparent widespread acceptance in developing countries, at elast one writer has argued that 'sustainable utilisation is bureaucratically and politically acceptable precisely because of its ambiguity'.

How has international law sought to circumvent the difficulties it faces in protecting the environment? It has developed different, often innovative strategies. One strategy is that of the 'umbrella' or 'framework' convention. This essentially means that states agree to a general set of principles for protecting the environment and then proceed to hammer out protocols with specific actions prescribed for them.

Another strategy is that of persuading states sharing natural resources to co-ordinate their municipal laws and establish common standards of utilisation of the shared natural resource.

A strategy to reconcile different interests of states, particularly developed and developing states, has been to give them different responsibilities. Thus in the Montreal Protocol, the need of developing states to benefit from the use of CFCs and their inability to produce or buy substitutes was recognised, and they were allowed to use CFEs longer than developed states; developed states were recognised as being the most responsible for causing ozone depletion, and more stringent and speedier CFE cuts were envisaged for them.

The ozone treaties, despite their innovative approach, also use an old weapon of international law to extract obedience from states—the sanction. In this case it is termed 'control of trade with non-parties' and constitutes Article 4 of the Montreal Protocol; it calls for banning the import of CFCs from non-parties; banning the export of CFCs to non-parties; curtailing the transfer of technology which might facilitate CFC production in non-parties etc. The sort of sanctions envisaged, however, reveal the importance of having the strong states backing the sanctions, and problems inevitably emerge when strong states themselves becomes the offenders.

International environmental law may also be strengthened if certain so-called 'emerging principles' become established as a state practice and hence customary international law. These include subjecting projects which may have a significant impact on the environment to environmental assessment, and the duty to inform all persons in a timely manner of activities which may significantly affect their use of a natural resource or their environment.

International environmental law is developing. It is flexible, and its utility cannot be denied. Yet it faces many problems as seen above. While it can solve some of these and sidestep others, it is surely not fair to expect it to tackle problems arising from the political organisation of the international system. And inasmuch as many questions to do with the environment such as development policy, state sovereignty etc., are political questions, it is surely the job of politicians to tackle these. The limitations of the law must be recognised. And arguably, the pressing need is for all those involved in the issues raised thus far, including economists, political scientists, sociologists, scientists etc., to join the environmental debate, and educate the politicians who finally determine international policy. (Mukund G. Rajan)

ENVIRONMENT PROTOCOL SHOULD NOT BE INEQUITIOUS

Global environmental cooperation is possible only when international agreements and protocols on the issue recognise the principles of equity and liability.

Even after the London 1990 amendments to the Protocol, the instrument remained inequitous and did not recognise the principle of liability.

Developed countries were trying to wrigle out of even this amended and inequitous Protocol. The financial mechanism should be re-examined. A dedicated fund was basic to the Protocol.

Technology must be accessible : Substitute technologies must not only be made available, but also made acceptable. The technology availability must be backed by funds to buy it to make it accessible, and there must not be any strings and conditionalities attached so that it is acceptable. Arguments about intellectual property rights and patents would be out of place here, for it is the responsibility of countries who are members of the Protocol to ensure that countries in a way that is acceptable to them.

The principle of liability of those nations who have been responsible for creating the problem in the first place has not been recognised. Thus funds are being made available by developed countries not simply because they have the money and the technology, but because of the principle of liability which accurately reflects the responsibility for the damage to the environment. This perception should establish a new path-breaking model for future international cooperation in the field of environment.

In the developing countries the use of CFCs (chlorofluorocarbons) was very small in per capita terms. Even after drastic reductions on the part of developed countries and assumed unrestricted use of CFCs on the part of developing countries, the former would continue to deplete the ozone layer at a rate substantially higher the developing world.

CHECKING DEGRADATION OF THE ENVIRONMENT THROUGH NATIONAL LAWS

The major merits of the Law on environment are: (1) claimants are to be paid for damages *per se* and they are not to plead and establish that death, injury or damage was due to any wrongful act, neglect or default of any person; (ii) compensation is to be paid for damages to property, people, flora, fauna, and for the cost of restoration of the quality of the environment including its eco-system and (iii) public interest groups are legally entitled to submit applications for liability compensation on behalf of the victims.

Hazardous Substances

The Law accepts the definition of 'hazardous substances' given in the Environment (Protection) Act, when exceeding such quantities as specified in the Public Liability Insurance (PLI) Act. It refers only to the non-living substances which can cause harm to human beings, other living organisms, property and environment.

Several living micro-organisms/genetically engineered organisms or cells as 'hazardous substances' are not included.

Biosafety Guidelines

Several public and private enterprises have set up R & D unity to absorb and adopt RDNA bio-technology for its commercial application in drugs, pharmaceuticals, agriculture, food products, brewery and other industrial applications, including bio-degradation and the restoration of the quality of the environment. The Department of Biotechnology, has already taken a commendable initiative and issued elaborate recombinant DNA bio-safety guidelines and regulations in view of the potential danger of biological agents as a means of mass destruction. These considerations call for designating harmful micro-organisms/ genetically engineered organisms or cells as 'hazardous substances' so that the victims of damage due to them or from products obtained through their use, can also claim liability compensations.

Environment Laws

The liability payments by the owners to individuals and compensation for damage to private property can be disbursed under the award made by the Tribunal. In cases of damage to the environment, such payments for the restoration of the eco-system are to be credited to the Environment Relief Fund established under the PLI Act.

Long Range Approach

The system of liability payments to the victims of pollution and other environmental damages is justified, but it is a short term approach. The long range approach ought to be to prevent environmental degradation at all stages, instead of waiting for an accident to occur to awaken us. This can be accomplished by (i) minimising all polluting processes and activities; (ii) undertaking continued remedial measures to slow down environment degradation; and (iii) formulating an effective system of interview, monitoring and implementation of bio-safety rules and regulations. In our conditions, it is necessary to provide for stringent punishments to personnel proved guilty of negligence in the discharge of their responsibilities of monitoring and investigating the offences under the Law. (K.L. Mehra)

GREENHOUSE INSURANCE POLICY AND PLAN

There has been growing concern that the accumulation of greenhouse gases in the Earth's atmosphere may lead to global warming. This concern has led to a number of proposals to set physical targets for limiting greenhouse gas emissions. With carbon dioxide (CO_2) believed to be responsible for over half of the human contribution to greenhouse gas emissions, the energy sector plays an important role in strategies to address potential climate change.

- Sensible greenhouse policy requires balancing benefits and costs. Fossil rules provide more than 90 per cent of the world's commercial energy. Before committing to a path that would need a major restructuring of the world's energy system, therefore decision makers must address two questions. What would reductions in emissions buy in terms of reduced environmental, damage? And what would be the price tag?

Neither question has a straight forward answer. At present, huge gaps remain in our understanding of the physical and biological processes that influence the climate system.

Uncertainty also persists regarding the costs of measures to limit emissions. Proponents of immediate controls argue that ample low-cost alternatives to carbon-intensive fuels are readily available. All that is needed, they say, is the political will to engineer the transition to a low-carbon economy.

Many economists, however, are less sanguine about the costs of emission abatement. If economically attractive options are available, they argue, what is preventing these alternatives from automatically entering the marketplace? They also fear that a rapid transition away from carbon-intensive fuels would be extremely costly.

These differing opinions present a dilemma for policy makers, since a case could be made for or against emission reductions. Addressing the issue is like deciding whether to purchase an insurance policy: If the cost is negligible and the risks are considered great, there is little reason not to buy. But if the insurance premium is expensive—that is, if there is a significant price tag attached to limiting the emission of man-made greenhouse gases—then it may be worthwhile to pursue alternatives to immediate cutbacks. These alternatives include an intensified commitment to research that could reduce climate uncertainty and research that could lead to the development of new energy supply and conservation technologies.

Buying Greenhouse "Insurance"

The whole greenhouse is fraught with uncertainty—about what climate changes might actually result from increasing carbon emissions, when these changes might occur, and what their effects might be.

The atmospheric models used to forecast climate have several important shortcomings, including poor spatial resolution, inadequate accounting for feedback mechanisms, and insufficient treatment of such factors as variations in solar output, volcanic activity, and Earth's reflectivity. Nevertheless, these models generally indicate that a doubling of atmospheric greenhouse gas concentrations from pre-Industrial Revolution levels—expected within the next century unless extraordinary measures are taken to reduce emissions—would increase average global temperatures by 1.5-4.5°C.

Possible consequences might *include changes in crop productivity around the world, migration of forest ecosystems, rise in sea level, and extinction of some endangered species.* Although a growing amount of research is being done, information about impacts is still at a rudimentary stage and the environmental, economic, and social consequences of global warming remain uncertain.

In such a context of great uncertainty and potential risk, there have been a variety of proposals to buy "insurance" against global warming by reducing greenhouse gas emissions as soon as possible. Nations committed themselves to adopt policies to control greenhouse gas emissions but setting no specific deadlines.

One policy mechanism often suggested as a way to bring about emission reductions is the carbon tax. Such a tax would be levied on the combustion of fuels according to their carbon emission intensity, in effect raising the cost of energy consumption to end-users. Coal would be particularly penalised, since it produces 21 per cent more CO_2 than oil per unit of energy consumption and 76 per cent more than natural gas. The size of the carbon tax that would be needed to bring about the specific emission reduction has been hotly debated, however, as have the cost and availability of non-carbon alternatives and the speed with which demand-side efficiency can be improved.

In the absence of a carbon tax or similar restrictions, in many countries coal would most likely maintain its dominant role, particularly in the non-electric energy sector, owing to the depletion of supplies of conventional oil and gas during the coming century.

A computer model of CO_2 energy-economy interactions developed, called 'Global 2100', can analyse the effects of specific carbon emission quotas for each of five major geopolitical regions and simulate their effects on regional economies. An energy technology assessment submodel makes it possible to evaluate the effects of introducing new technologies, including the effects of the speed at which they replace older, more carbon intensive technologies. Using 'Global 2100', the researchers have also explored the costs associated with various carbon tax and technology forcing strategies that might be adopted everywhere.

Much of their recent work has focussed on assessing the costs of the so-called global stabilisation scenario. This scenario assumes stabilisation of global emissions at 1990 levels. Richels notes that "there is clearly no unique formula for achieving global stabilisation of emissions. Most proposals recognise the need for some growth in CO_2 emissions in developing countries, where per capita emissions are very low in comparison with developed countries.

One widely discussed proposal involves having developed countries cutback emissions by 20 per cent and letting levels developing countries increase modestly—which would result in net global emission stabilisation.

According to the analysis, GDP losses for industrialised countries following the 20 per cent reduction scenario would exceed one per cent, but many developing countries would ultimately be more severely affected economically even if they were allowed a modest increase in their carbon emissions over the next century.

Manne and Richels also examined the size of the carbon tax that would be necessary to bring about emission changes of this magnitude. The purpose of the carbon tax would be to raise the price of fossil fuels sufficiently that consumers would switch to less carbon-intensive alternatives.

The Manne-Richels analysis is based on what the authors believe to be realistic assumptions improvements in the energy sector during the coming decades. However, they stress that technology forecasting is a tricky business, and they show the sensitivity of their findings to alternative sets of assumptions.

A major reason that the costs of reducing carbon emissions are likely to be high is that the target date set for achieving a stable 20 per cent reduction is 2010 — a short time compared with either the rate of major energy facilities are replaced.

R & D

Technological development continues to provide one of the bright spots in the analysis of the greenhouse issue. Compared with the cost of either a carbon constraint or the potential impacts of global warming, the cost of the R&D needed to produce new energy technologies is small indeed. Without trying to specify which technologies are most likely to succeed, the Manne-Richels analysis dramatically illustrates how important their overall contribution could be.

On the demand side, technological improvements offer the promise of higher efficiency and a further decoupling of economic growth from energy consumption. Manne and Richels found that further investment in this area could have an enormous pay off. For example, doubling the rate of exogenous (not price-induced) efficiency improvements could nearly halve the costs of a carbon constraint.

The biggest long-term pay off will come from successful R&D on the supply side. As crude oil reserves are depleted, coal-based synthetic fuels will become competitive. But such synthetic substitutes produce twice as much CO_2 per unit of useful energy as oil and would thus be penalised severely by a carbon constraint.

The development of lower-cost renewable fuels, such as ethanol produced from biomass, could provide a noncarbon alternative, but serious questions remain about the size of their potential contribution because of such factors as competing uses for land and water resources. The possible contribution of renewable fuels in the carbon-constrained of renewable fuels in the carbon-constrained analysis was assumed to be 10 quads per year. If the contribution from low-cost renewables could be doubled, however, the cumulative cost of reducing carbon emissions could be lowered by more than one-third.

Fortunately, there are a number of promising carbon-free electricity-generating alternatives. Among these new technologies are direct solar conversion, wind, advanced nuclear reactors, and biomass conversion. Without trying to predict the contributions of specific technologies, Manne and Richels estimate that if *the cost of electricity from carbon-free alternatives could be made competitive with that of electricity from new coal-fired power plants, the cost of a carbon constraint could be reduced by more than one-third.*

In their analysis, Manne and Richels assume that environmental concerns will eventually stimulate the development of economically competitive alternatives to coal-fired power plants. As a result, they project that the electricity sector will gradually become less carbon-intensive even in the absence of measures to limit CO_2 emissions.

This is not the case for the nonelectric energy sector. *When conventional oil resources are exhausted, they are all likely to be replaced with even more carbon-intensive synthetic fuels.* This means that — *in the absence of measures limit CO_2 emissions the nonelectric sector is likely to become more carbon-intensive over time.*

What this tells us is that if we are really serious about bringing down total carbon emissions, electrifications should be an important part of the strategy. Electric and non-electric energy can be substituted for one another in many markets. Electric heat pumps, for example, offer an alternative to residential oil burners. Electric vehicles offer an alternative to the internal combustion engine. Such interfuel substitution could lower carbon emissions. But economically competitive, long-term alternatives to coal fired power generation must be available in sufficient quantities to make a difference. Again timing is important — the transition to a more electricity-intensive energy economy based on carbon free generation cannot be accomplished over-night. (John Douglas)

ENVIRONMENTAL COSTS

"Global environment facility (GEF) is being restructured". A restructured GEF, democratic and transparent in its functioning, would be shortly in place. However, there is (as yet) no sign of even the 20 per cent share of the industrialised countries towards funds for implementing and providing adequate environmental safeguards and incorporating environmental costs (i.e. costs incidental to the prevention of environmental damage) in the trading prices of several items traditionally exported to the North countries over the years earning valuable foreign exchange but leaving in bad shape the immediate neighbourhood of the manufacturing facility in the absence of suitable mechanisms for monitoring and treatment of the industrial waste prior to its disposal.

The leather industry earning valuable foreign exchange are slated to grow in a big way is a typical example that will serve to bring home the harsh realities. The pollution of underground water and loss of fertility of soil are the twin aspects of the environmental degradation brought about by the open disposal of untreated tannery effluent. The pollution control Boards have not been equal to the task of making the leather industry comply with the stipulations of pollution control largely due to the fact that there are a large number of small units widely dispersed in a given region and burden of cost incidental to effluent treatment is something that these small units cannot bear individually. As of now the problem has remained intractable, one can expect a similar situation to exist in other developing countries as well-exporting products and earning, direly needed foreign exchange at the expense of their own environment.

The only way for the developing countries to be weaned from their path of peril to environment is to ensure for them a fair trade price for producing environment-friendly products. And this will come about only if the countries get together and arrive at mutually acceptable guidelines for the incorporation of environmental costs in the trading price of exports items while ensuring the adoption of international standards in the schemes for environmental protection. (N.K. Suryanarayanan)

III

POLLUTION CONTROL IN SOME SPECIFIC INDUSTRIES

In Pulp and Paper Industry, as in most process industries, there are three kinds of pollution, viz., Water Pollution, Odour Pollution and Solid Wastes.

Water Pollution

The Pulp and Paper Mill effluent generally contains:

— Suspended solids including bark particles, fibres, filler materials, etc.;

— Dissolved organic materials including lignin; compounds, carbonhydrates, etc., causing BOD load;

— Dissolved inorganics on account of pulping and bleaching and paper making chemicals; and

— Coloured bodies — lignin compounds and dyes.

In-Plant Effluent Control

The most obvious pollution control technique is the proper operation of a mill. This means avoidance of spillages and leakages and optimum capacity utilisation of the equipment. Because of the complex chemical reactions that occur during pulping and bleaching processes, over loading or improper maintenance of plant and machinery can lead to disproportionately high levels of pollution. Greater water recycling can also significantly reduce pollution while increasing production and saving valuable fibres, chemicals and heat.

Effluent Treatment

The primary treatment removes about 80-85% suspended solids and reduces BOD upto 15-20%. Secondary treatment is required for further reduction in BOD load. Among the different secondary biological treatments available, the Activated Sludge Process is the one most commonly used as the degree of BOD removal is the maximum, 80-90%.

In some places, the effluent, after primary treatment, is used for irrigating crops like sugar cane, wheat, banana, etc.

As the conventional aerobic activated sludge process for the secondary treatment of the effluent involves extremely high operating costs, especially with regard to energy, Pulp and Paper Industry is now looking for newer technologies like anaerobic digestion of the effluent. This facilitates recovery of energy from waste in the form of bio-gas.

Even after the secondary treatment, the effluent still poses the colour problem.

Odour Pollution and Control

Most odour pollution problems with Pulp and Paper Industry are in the Kraft (Sulphate) process and, to a lesser extent, Sulphite process. The atmospheric emissions include both particle and gaseous matter.

The particulate matter emissions are from the chippers/choppers, bamboo and wood dust, coal particles from coal crushers, sodium sulphate and sodium carbonate particles from recovery furnace, lime dust, etc.

Paper industry is taking care of particulate matter by *installing suitable dust particulate collectors like mechanical dust collectors, electrostatic precipitators (ESP), bag filters, etc.* Practical experience has shown that the ESP can effectively control particulate emissions from Recovery Boiler and other producers of dust-laden gas in the Pulp and Paper Industry.

Elimination of odour is not as straight forward as the elimination of particulates and/or odourless gases. Odours are difficult to measure and a tolerable or acceptable level is not easily defined. The odour threshold levels for all major reduced sulphur compounds from the pulping process are very, very low. They are of the order of one part per billion parts, i.e., one microgram per cubic meter.

Some of the odour control techniques available are:

— Oxidation of black liquor to avoid the volatilisation of sulphur compounds during evaporation;

— Scrubbing of digester gases with alkali/chlorine;

— Collection and burning of the digester gases either in a separate incinerator or in the rotary lime kiln.

Solid Wastes

The major solid waste materials that generate from Pulp and Paper Industry are the wood/bamboo dust, bagasse pith, coal dust, cinder, line dust, lime sludge from causticization plant and effluent sludge.

The increasing quantity of solid waste is a serious environmental problem not only in developed countries but also in developing countries. Until recently, disposal was the only technical and economical option. Of late, the recovery from waste has attracted a growing interest in view of the increasing awareness about conservation of natural resources. Wood/bamboo dust, bagasse pith and effluent sludge and nowadays used, mixed with coal, in modern multi-fuel biolers. Flyash is finding increasing application in making construction materials.

Lime sludge, which is used as land fill at present, can be converted to lime in Rotary Kilns. But, due to high silica content in lime sludge, most of the Mills are unable to burn the lime sludge for the regeneration of lime.

Government has enacted (a) Water (Control and Prevention) Act, 1974; (b) Air (Control and Prevention) Act, 1981 and (c) Environment Protection Act, 1986 to control land prevent pollution of water and air. Pollution Control Boards have been set up at the Centre and in the States as monitoring agencies and they have prescribed certain standards for the discharge of pollutants.

Enactment of fool-proof laws and pragmatic minimum national standards (MINAS) and provision for appropriate incentives are, indeed, steps in the right direction. Equally important, if not more, perhaps is availability of environmentally friendly technologies and products such as industrial safety and pollution abatement equipment, monitoring and control systems. Also, attention should be focussed not only on laws enacted, standards laid down and incentives offered so far but also on gaps in the availability of appropriate technologies, equipment and control systems.

GLOBAL TRENDS

Advanced Waste Water Treatment

Some of the advanced waste water treatment methods used for the removal of long chain biodegradable organics as a polishing step are: *Absorption by granular activates, carbon, Reverse Osmosis for the removal of dissolved and suspended solids by almost complete membrance rejection of impurities in waste water, Ion exchange etc.*

New Technologies

Not too long ago, the primary concerns were BOD (Biological Oxygen Demand and TSS (Total Suspended Solids). Traces of dioxin (2, 3, 7, 8 tetra chlorodibenzo-p-dioxin) found in effluent, results in major environmental hazards.

Process technologies for pollution reduction: These are technologies like Modified Cooking Process, Oxygen Delignification, Modifed Cooking, Continuous Digesters, Extended Delignification (Rapid Displacement Heating System), Improved Brown stock washing and modified bleaching systems like Chlorine-dioxide substitution in the initial chlorination stage, counter current washing in the bleach plant, Oxidative extraction using oxygen and/or hydrogen peroxide, especially to reduce colour of the effluent as extraction stage is the major contributor of bleach plant effluent colour and burning of sludge to minimise the amounts going to landfill besides low odour Recovery Boilers.

These technologies would provide two significant benefits viz., *reduced hazardous pollution and greatly improved production efficiency.*

Of late, a few more technologies *like Biotechnology and Organosol pulping using solvents like methanol bleaching with Ozone, etc.* are being explored especially to *meet the environmental pressures.* Last year, mill-scale enzymatic pre-bleaching trials were completed at Anaekoskikraft mill in Finland. Substantial reduction in consumption of chlorine and Absorbable Organic Halogens (AOX) levels were reported.

A modern bleach kraft mill using the latest technology will produce effluent in which the dioxin like methanol will be non-detectable. Many countries such as Sweden are taking steps to control total organic chlorides (TOCI) and absorbable organic halogens (AOX) for further minimising dioxin levels in the effluent.

With the availability of right quality metallurgy, it has been possible to install bleach plant with ''zero-discharge'' by using Rapson-Reeve process.

The enormous strides taken by the developed countries like USA could be gauged from the impressive reduction achieved by an integrated Kraft mill between 1964 and 1990 in the matter of effluent discharge (85%), TSS waste load (75%) and BOD load (70%) per ton of bleached pulp.

In Pulp and Paper Industry today, with the state-of-the-art technology and equipment, waste water can be cleaned of over 90% of its solid waste load and over 85% of the dissolved solids and chemicals washed from the pulping and bleaching operations.

Even discolouring matter in the effluent can be greatly reduced by physical and chemical types of treatment. *Before Mill effluent is discharged into public waterways, it can be restored to non-toxic and aerated condition.*

With state-of-the-art control, over 99% of particulate emissions and over 95% of gaseous contaminants may be minimised.

The mills are to be modernised and have to adopt newer bleaching technologies and micro-processor control operations of paper making. Besides, all the mills use conventional CEHH sequence for bleaching because of its lower cost and as it is not practical, at the moment to reuse chlorination stage back water, which alone accounts for 40% of the effluent loads of the mills, due to its acidic nature and also non-availability of the right quality stainless steel at reasonable prices. Installation of CEDED sequence bleaching process is more efficient and should be adopted.

The pulp and paper industry uses large quantities of water during the manufacture and also generates much larger quantity of effluents than any other industry. The waste water generated from big mills varies from 400 to 500 cubic metres per tonne of paper produced. Hitherto the easiest way to deal with this large effluent quantities was to site the paper mills astride rivers so that the effluents from the paper industry are let into the river streams, where they get diluted with the river flow. This practice has been causing considerable ecological damage. Awareness has been created to control this pollution.

The paper industry mainly generates two streams of effluents. The first one is black liquor from the pulping section. The second one is from the paper machine section which is called white water. While the white water can be effectively treated and clarified and let-off into the fields without presenting much ecological damage, black liquor has been causing considerable concern allround for its disposal owing to its high lignin content.

There are several methods used in the pulping process to release the cellulosic fibres for paper manufacture like kraft process, sulphite process and soda process. Whatever be the process adopted, there are vast quantities of black liquor from the pulping section to be dealt with and means of dealing effectively with this objectionable effluent have to be found out.

POLLUTION CONTROL SYSTEM IN A TYPICAL/MODEL PHARMACEUTICAL FACTORY

Main Features

* The system of pollution control is not an afterthought but has been planned integrally with the production process.
* All waste streams have been segregated for more effective treatment.
* The unit has inbuilt monitoring and auditing facilities.
* No liquid effluents go out of the factory. Either they are incinerated or bio-treated, filtered and reused within the factory land for gardening.
* Power supply to the production process may be interrupted but not to the pollution control devices which are connected to diesel generating sets.

* Nitrogen blanketing has been provided around solvent and waste tanks to prevent evaporation and associated air pollution.
* Accidental spills are taken care of with dykes and catch its and ceramic diffuses are provided to prevent solids escaping through process drains.
* Consumption of raw materials and natural resources (fresh water, solvents, oil, gas, power, etc.) is monitored and recovery systems have been instituted to reduce waste and the pollutant load.

Basically, the effluents from the process consist of washings and gaseous emissions. The highly salt laden washing are taken to bitumenlined shallow ponds where solar heat is allowed to evaporate the water and the residual salt is trucked out. These solar ponds can handle up to 500 litres of effluent per day.

The biodegradable washing are neutralised and sent to specially designed aerobic reactors where microbes play a major role in oxidising and degrading the organic and carbonaceous matter. This bioreactor can handle about 60,000 litres of waste water a day. The clear water from the reactor is used for gardening, thereby saving consumption of some 15,000 litres a day of fresh water.

The healthy growth of the ornamental bushes and trees in the factory testify to the effectiveness of the bio-treatment. The acidic gaseous emissions from the process reactors are scrubbed in a packed column with a counter current of caustic for neutralisation. The washings from here are led to the solar ponds and the scrubbed gases, conforming to the Pollution Control Board standards, are let off into the atmosphere.

Sophisticated Incinerator

The piece-de-resistance of the pollution control system is the incineration plant. A highly sophisticated unit this incinerator can dispose of five tonnes a day of liquid waste. It is designed for destruction of organic residuals and gaseous wastes like the highly odorous mercaptans by burning them at 1200°C. This incinerator boasts of a programmable burner operation system, an ultraviolet optical flame sensor, proportional, temperature controllers, a multifuel burner and atomising injectors for burning low calorific value liquid wastes etc.

On top of the incinerator, there is a wind speed sensor and a wind direction indicator, both of which are coupled to an automatic data logger. With the wind rose patterns available through these readings, it will be possible to map out the spread of the stack emissions from the incinerator in the surrounding countryside as well as at ground level in the factory environs.

The integrated pollution control system, particularly the incinerator, will enable the plant to be highly flexible in taking up the manufacture of new products.

Not much time will have to be spent in devising ways of treating the effluents from the new product lines. The factory has on the cards a programme to take up the manufacture of quinolones in the next few months.

The factory has extended a helping hand to assist other factories in the area in treating their wastes, since there is considerable spare capacity in the incinerator. The company is also offering the services of the sophisticated environmental laboratory for testing the samples of wastes of units. The company has also offered to share the wind data collected by it with the Met. Department and other regulatory bodies.

SPINNING OFF HEAT WITH LESS POLLUTION

Like a miniature tornado, the tunnel-shapped flame blazing in an experimental burner rotates furiously, spun by precisely aimed jets of air. The burner's design can substantially reduce the amount of polluting nitrogen oxides emitted by utilities and industry as they produce power and run manufacturing processes.

When virtually anything burns, gaseous nitrogen which makes up 79 per cent of the atmosphere, mixes with the oxygen in the burning mixture to generate nitrogen oxides (NO_2). Nitrogen containing fuel such as coal and oil produce even more NO_x. And in sunlight, the NO_x can mix with hydrocarbons from, say, automobile fuel to form ground-level ozone, which causes respiratory stress and other health problems. Nitrogen oxides are also components of acid rain.

In conventional burners, fuel and air, mix vigorously from the start. This makes for efficient combustion but also provides the right conditions — high temperatures and plenty of oxygen for NO_x formation. The radially stratified ''flame core burner'' delays this mixing and thus limits the formation of nitrogen oxides.

In the first stage, the rotation of the cylindrically shaped flame pushes the relatively cool, dense air simultaneously piped into the combustion chamber to the outside of the flame, almost completely separating the fuel from the oxygen. Inside the heat of 1,200-1,500°C exists a calm zone, much like the eye of a hurricance, where the fuel pyrolyses, or breaks apart chemical bonds in the absence of oxygen. This creates a waste product of inert nitrogen gas along with heat.

About three feet beyond the fuel nozzle, the flame "blossoms out" and collapses into a doughnut-like shape. In this second stage, air mixes with the pyrolysed fuel, which ensures that the fuel is burnt completely. Carbon dioxide and water are the primary waste products.

The experimental burner can use natural gas, oil, or pulverised coal as a fuel with only minor ajdustments. With natural gas, the burner emits as little as 14 to 16 parts per million nitrogen oxides, a significant reduction from the 200 ppm generated in a conventional burner. Burning oil yields about 80 ppm compared with as much as 600 ppm otherwise. Combustion-modification techniques like those in the radially stratified burner will most likely give the best bang for the buck.

FAST WAY TO METHANOL

Unexploited reserves of natural gas in remote locations could be much easier to tap in the next century, thanks to breakthroughs in catalysis. The systems should enable gas engineers at well sites to convert methane in the natural gas into methanol, a liquid which is much easier, safer and cheaper to pipe to consumers hundreds of kilometres away.

It's so costly to ship it (methane) at present because it's so expensive to condense it with high pressures or to liquify it by cooling it. Methane is induced into methanol with a mercury catalyst at the relatively low temperature of 180°C.

Existing methods of converting methane to methanol react gases together at temperatures of about 900°C. They also require large energy hungry equipment, so are impractical for remote or inhospitable locations, and convert less than five per cent of the methane in the natural gas into methanol. Taube's method converts some 43 per cent, with the rest of the methane mostly forming carbon dioxide.

The formation of methanol is a "halfway house" on the way to carbon dioxide, and hard to make reaction stop there. A mercury catalyst will promote the reaction of methane into methanol, but it is then very difficult to stop it continuing and turning the methanol into carbon dioxide.

The new process carries out the reaction in sulphuric acid rather than water. This converts the methane into methyl bisulphate, and because a mercury catalyst is not very good at reacting methyl sulphate any further, it simply keeps working on the methane. "The mercury sees either methyl bisulphate or methane, and it would rather react with the methane by a factor of 10,000. This selectivity for methane is one key to the process.

In the next stage, the methyl bisulphate is passed to another vessel and dissolved in water, which leads to the formation of methanol. Sulphur dioxide gas bubbles off and it could be recycled to reform sulphuric acid for the first stage. The methanol can simply be distilled off and piped to consumers.

A new method for making methanol is mixture of methane and oxygen through a hard, sponge-like catalyst which has its surfaces covered with the metal rhodium.

The reactants are converted into a mixture of products, including carbon monoxide, hydrogen, water and carbon dioxide. Steam added in the next stage changes the mixture into a hydrogen and carbon monoxide mixture called 'Syngas'. The Syngas can be easily converted into methanol with a catalyst of copper and zinc oxide.

IV
RECYCLING AND REUSE OF WASTE AND BY-PRODUCTS

Recycling Consumer Waste and Packaging Materials

The consumer item and more so the packaging, in their making or disposing affect the environment. For example, use of wooden crates means cutting trees for making them and hence they are not environment-friendly. Similar is the case of paper which is made from wood pulp. But if it is recycled, then the quantum of the total raw material required is reduced and hence it is a little more "environment-friendly".

When buying products it is examined whether they can be reused or if they cannot then they should be easily biodegradable. This means that the material can be broken down by the bacteria in the environment. This normally occurs in well-designed compost operations. In general, biodegradable products are preferable for the items that cannot be recycled.

It is difficult to remelt plastics and reuse them as they are not biodegradable. Therefore it becomes a headache for environment.

Recycled Paper : Making just one tonne of paper uses the equivalent 17 trees. Every year Americans use more than 67 million tonnes of paper and it is the largest single component of the solid waste stream. Paper products now consume 35 per cent of the world's annual commercial wood harvest, and are expected to consume 50 per cent by the year 2000. Everyday American business generate enough paper to circle the earth 20 times.

Fortunately paper is recyclable. The recycled paper can be made from waste paper instead of virgin wood pulp. First it is mixed with water in a hydrapulper, which is like an industrial sized kitchen blender. It mixes the paper and water, pulling inks away from the paper fibres and separating the fibres themselves. De-inking chemicals are also added at this time.

The mixture is further cleaned and refined as it passes through several different-sized screens. The screens separate the fibres from contaminants such as paper clips and staples. Once the mixture is cleaned, the clean pulps is re-formed — pressed into sheets, dried, finished and placed on rolls and made into new paper products. In most cases the clean pulp is mixed with wood pulps (virgin fibre), producing paper with partial recycled fibre content.

Waste paper is divided into "grades". High grade papers are office paper, stationery or computer printout paper. Lower grade papers are newspapers and telephone directory paper. Corrugated boxes, paper board and craft papers are also recyclable. But each type can be recycled not into the next grade but in the same grade. For example newspapers cannot be recycled into office paper, but can be recycled into newsprint and telephone directories.

There is the "post-consumer waste," meaning waste paper collected after use from offices, homes etc., as well as any material that is normally disposed in landfills. The higher the post-consumer content of a recycled product, the more beneficial it is to the environment. On the other hand, "Pre-consumer waste" is known as manufacturing waste. This is waste produced by converting or printing operations and includes any scrapes, trimmings, overruns etc. Most recycled products contain a portion of pre-consumer and post-consumer waste.

U.S. specifies use of at least 50 per cent waste paper content for printing and writing grades. This includes stationery, office paper, copier paper, envelopes, book paper, bond paper, ledger, cover duplicator and offset printing paper.

Recycling Project: In the U.S. an innovative national office paper recycling project has been started for office paper recycling. The goal of the project is to maximise the recycling of office waste paper and to minimise its disposal. It is a collaborative effort by corporations which manufacture paper, produce machines that put images on paper and those that collect and recycle office waste paper. In addition the project includes public interest groups that represent state and local governments which are interested in handling the solid waste crisis. They have come up with a comprehensive guide for buying recycled paper products.

Packaging Materials: In addition to recycling there is a concerted attempt to reduce waste since it helps to conserve resources and save the environment. These include buying recycled products with less packaging. More and more thought is going into the quantum of packaging materials used. A case in point is the attempt by companies manufacturing electric pumps and motors who have applied analytical techniques like value engineering to make window type crates instead of solid ones. They can also be used continuously instead of abandoning them after a single use by circulating it to various points.

Countries like Germany have specified that the packaging of products which are exported from India to them and other countries should conform to the standards prescribed by them, to enable them to recycle them easily. For example, they speedify the chemicals that are banned the printing ink which is used on the crates which cannot be processed according to their recycling operations.

The packaging materials themselves are classified as having 'high or low sympathy' to the environment. For example plastic boxes have 'low sympathy' and paper and metal which is easily recyclable has 'high sympathy'.

An aggressive environmental management really works in a preventive manner. The priorities are (a) use of substitutes to enable recycling, (b) design for recycling, (c) economically recycle, (d) reduce the consumption itself, (e) sorting of waste so that recycling can be done in an appropriate manner for the different components of waste and (f) dispose after making sure that the waste is biodegradable.

The Indian Scene: Companies who are exporting to EEC countries are now finding themselves in a quandary particularly in view of the stringent specifications of packaging materials apart from the product specification which should conform to ISO 9000 standards. Companies like Bosch have made it clear unless the packaging strictly conforms to their laid out specifications it will have to be taken back to the home country at the exporters cost! They have their own recycling processes and standards and the material used in export packaging here should be amenable to these treatments.

Examples of environment-friendly initiatives, are the use of bagasse which is agricultural waste for paper making instead of wood pulp, also energy conservation measures and recycling within the plant.

WASTE RECOVERY — EXTRACTION OF GOLD

Biotechnology can help enhance gold recovery from refractory ores and waste tailings. In the next decade, this technology is set to revolutionise gold production, making it economically viable to recover precious metals even from wastes and lean ore deposits.

The current Indian practice of gold extraction is a century old technology — involving fine grinding, cyanidation and gold cementation by zinc. This process, originally developed for free milling relatively high grade ores, is inefficient with leaner and refractory ores containing less than about five grams of gold per tonne of ore. In this context, biomineral technology offers cost effective and environmentally safe techniques which could convert resources classified unviable into potential treasure chests.

Application

Biotechnology could be effectively applied to recover gold from the following unconventional and waste resources :

- Tailing dumps accumulated at mine sites over the years amounting to several millions of tonnes, which in a few locations contain one gram of gold per tonne of ore.
- Lean grade free milling ores containing less than 2-3 grams of gold which cannot be economically processed.
- Refractory ores — sulphidic or carbonaceous, where finely disseminated gold particles are locked-up, making them unsuitable for direct cyanidation.

The following aspects of bioprocessing are relevant to gold metallurgy: benefication (enrichment) of low grade and high sulphidic gold-bearing ores; biooxidation of gold-bearing sulphide minerals to liberate encapsulated gold; direct solubilisation of gold; recovery of gold and associated base metals from cyanide effluents through biosorption and bioaccumulation and biodegradation of cyanide wastes for environmental pollution control.

The applicability of biomineral technology thus extends over the entire spectrum of gold production from mining to waste disposal.

Several microorganisms are involved in the above processes and one has to choose appropriate organisms for any desired result. Another unique aspect is that all of them are present in the deposits and mine wasters ubiquitously as an essential component of the mine ecosystem.

Mineral Benefication

Several types of autotrophic and heterotrophic bacteria aid in ore enrichment. Floatation is the universally adopted benefication technique for sulphide ores. Gold-bearing sulphides containing pyrite, pyrrhotite and arsenopyrite are often floated to yield a gold-rich concentrate before leaching.

Conditioning the finely ground ore with Thiobacillus ferrooxidans prior to floatation modify the surface properties of the sulphide minerals so as to enhance their recovery in the concentrate. Since the gold values in the refractory sulphide ores are locked-up in the sulphides, microbially aided floatation eliminate the need for handling large tonnages of low-grade materials. Sulphate reducing organisms of the type Desulfovibrio aid in the production of floatation modifiers such as sodium sulphide and hydrogen sulphide. Similarly, heterotrophs including fungiproduce organic acids and surface active agents which aid floatation. By controlled bacterial conditioning, the desired mineral in an ore can be selectively enriched. Also there is no need for costly reagents.

Refractory Ores

Finely disseminated, gold encapsulated in a sulphide mineral such as pyrite and arsenopyrite does not readily respond to cyanidation and hence they are termed refractory. Prior biotreatment of such ores or their floatation concentrates would readily dissolve away the sulphides, enhancing gold recovery. The bacterium, Thiobacillus ferrooxidans present in situ in the deposits, selectively oxidise the sulphide minerals, resulting in the formation of acidic ferricsulphate solubilising the arsenic and other base metal constituents.

Direct cyanidation of these ores without prior biotreatment would result in a gold recovery of upto 90 per cent. Biooxidation is carried out in specially designed and controlled bioreactors for even up to five days. An industrial plant is in operation in South Africa since 1986 and bioleaching facilities are being commissioned in a few other countries.

Heap Leaching

This method is suited for lean grade ores and plant tailings even 0.2 grams of gold per tonne of ore. The low grade ores are heaped on impervious platforms and the bacteria are allowed to multiply by providing appropriate conditions. Sprinkling sulphuric acid solution (pH two) along with essential nutrients enable bacterial growth and consequent oxidation of associated sulphides.

Tailing dumps, pose solution percolation problems due to fineness. The compact mass need be loosened by ploughing to let in air, acid and nutrients. The leaching cold be carried out in stage in layers also.

In both cases, the heaps should remain under prescribed condition over a period of, say, a few weeks or even months. Then the bioreacted mass is subjected to cyanidation for gold recovery.

Heterotrophic bacteria Bacillus sp. Bacillus megaterium, Bacillus mesentericous, and Bacillus subtilis solubilise gold from ores by their innate capacity. The ionised gold is complexed by the microbial metabolites. Mycelium of the fungus, Aspergilus niger and wastes from agro and food industries are effective gold solubilisers.

Direct gold solubilisation is carried out at an alkaline pH of 8-9. Ninety per cent of gold has been solubilised in 10 days inside the laboratory (so far no commercial process has been reported). This method does was with the costly and toxic cyanidation step.

Bioaccumulation is the ability of microorganisms to absorb and accumulate metal ions from aqueous solutions. High metal sorption capacities have been reported for various bacteria, fungi, yeasts and algae. The biosorption of Aspergillus sp. biomass exceeds that of even activated carbon by 10 to 12 times in case of colloidal solutions. Gold concentration coefficient in algae is ten thousand times more compared to that of a sea water.

Some of the fungal strains Aspergillus niger, Penicillium citrimum and Rhizopus arrhizus also absorb significant qualities of gold, zinc and iron from cyanide effluents. Immobilised cells of yeast. Saccharomyces cerevisiae are efficient in copper accumulation from cyanide solutions.

The fungal strains possess certain unique attributes. They exhibit a filamentous or hyphal growth providing a larger biomass surface area for adsorption. Their cell walls contain chitin, a natural polymer known for its ability to complex gold and other heavy metal ions.

Gold extraction plants are faced with a major environmental problem in disposing the massive cyanide containing wastes. Bacteria can convert and/or metabolise cyanide species to inert compounds and simultaneously fix the heavy toxic metals in a harmless state. The bacterium, pseudomonas sp. can almost completely detoxify free cyanide, thiocyanate and metal complexed cyanides. The Homestake gold mining company in the U.S. has recently commercing company in the U.S. has recently species and bioadsorption of toxic heavy metals.

Many cyanide-tolerant fungi such as Fusarium Rhizopus and Aspergillus and bacteria such as Bacilli. Actinomyces and Pseudomonas are naturally occurring and can be grown on a massive scale at mine sites. Compared to the other alternatives such as alkaline chlorination and peroxide treatment, the biological processing is competitive, and flexible for an integrated process circuit. Extracting gold from low-grade ores and tailing dumps (containing less than two grams of gold per tonne of ore) is a tricky business. Several bacteria and fungi are good at it and offer a economically viable alternative. (K.A. Natarajan)

REUSE OF SPENT CATALYST

Japanese developed a technique to recover and refine at low cost useful metals such as vanadium and nickel from catalyst exhausted from refineries. The recovered metals can be used for pigments or catalysts. The company plans to start a recycling business to process all of the used catalysts exhausted from Japanese refineries.

The desulphurisation catalyst used by refineries consists of alumina as the carrier impregnated with active metals. Molybdenum is one of the active metals on which vanadium present in crude oil is deposited in the desulphurisation process.

The technique recovers molybdenum and vanadium separately from the used catalyst. The used catalyst is first fired at 750°C or higher to remove oil and carbonaceous matter. It is then treated with caustic soda to extract vanadium and molybdenum in the form of sodium vanadate and molybdate.

The mixes solution is concentrated and cooled to separate them by solubility difference. Each solution is adjusted at a given pH to recover molybdate or vanadate, which are approximately 99 per cent pure. About 100 kg of each salt can be produced from one tonne of the spent catalyst.

Alumina as the residue can be used as the filler for gravel. Recovered molybdenum and vanadium are competitive cost-wise because of a simplified recovery process which does not need high temperature or pressure. In addition to catalysts, they can be used for pigments, colouring agents, and steel.

METALS FROM WASTE

Treatment of industrial wastes for the recovery of metals has been developed through leaching, solvent extraction purification and electrowining.

Since most of the industrial wastes like brass dross, zinc dross, mixed scrap, spent catalysts grinding dust, flue dust, sludge and slags are rich in metal values ranging from 20 to 90 per cent processing through leaching.

Purification and electrowining steps are economical, asserts the statement. The hydrometallurgical routes is preferred to other techniques because of its simplicity convenience and product quality.

JAPANESE RECYCLE SPENT FUEL — SPENT FUEL REPROCESSING PLANT

The Japanese have constructed their first commercial plant for reprocessing spent nuclear fuel, which constitutes a major part of Japan's plan for recycling the fuel used in its nuclear power plants.

In the cycle, spent fuel from commercial nuclear reactors is chemically treated to extract the unspent uranium and plutonium produced, which are then expected to be recycled as fuel in the nuclear power reactors.

Spent fuel produced being stored in pools at the power plants. The stockpile of the spent fuel is expected to decrease when the cycle goes into operation. The reprocessing plant is expected to produce around five tonnes of plutonium from spent nuclear fuel annually.

Japan has a prototypes fast breeder reactor and also plans to use a mixture of uranium and plutonium as fuel in ordinary reactors. But even when spent fuel is reprocessed into plutonium, some highly radioactive wastes will remain and safe way to store them has to be worked out.

Japan's heavy reliance on plutonium has come under the spotlight while other nations are abandoning their original plans to build plutonium-operated fast breeder reactors. Because the reprocessing plant in Japan is not yet complete, Japan has contracts with French and U.K. companies to reprocess spent fuel into plutonium.

A cargo of plutonium has reached Japan from a reprocessing plant in France. The plutonium shipment was under fire from anti-nuclear groups such as Greenpeace and from governments along the shipment route because of the highly toxic nature, of plutonium. In addition to plutonium's cancer-causing nature some fear the possibility that plutonium will be used for nuclear weapons.

DISPOSING SULPHUR WASTE

Offensive sulphur wastes from sulphuric acid plants can be used to make sulphur-coated urea granules for use as fertilizers.

This may solve the problem of disposal of hazardous sulphur wastes besides yielding high-grade sulphur-coated urea, which helps control the release of nutrient to suit the individual needs of crops.

Almost seven tonnes of sulphur wastes are discharged for every tonne of sulphuric acid manufactured, scientists estimate.

First sulphur is separated from the wastes, using a suitable solvent such as carbon disulphide. By treating fertilizer-grade urea with the sulphur solution and evaporating the mixture, sulphur-coated urea granules.

RECYCLING TYRES

Landfill is till the most popular method of disposing of waste tyres but recent efforts have concentrated on burning the tyres for electricity generation. There is a ready market for rubber granules or crumb.

Rubber granules are used in making carpet underlay, car bumpers and other products.

Tyres are only 60 per cent rubber, the remaining 40 per cent is made up of roughly equal amounts of steel and textile fibres. Tyre plant first shreds the waste tyres into lengths of around one metre — the shredder can cope with all sizes of tyres from tractor tyres to small car tyres.

One granulator then cuts the shredded tyres further and the metal is separated by a magnetic separator. A second granulator reduces the size even more and the fibres enter an air separator. Finally rubber crumbs of about seven mm in size are produced for use as a raw material of carpet underlay, car mud flaps and bumpers and playground safety surfaces. More than 80 per cent of the tyre — the rubber and the steel can be recycled.

BIOGAS FROM SALVINIA WEEDS

The nuisance aquatic weed Salvinia, the world's most intransigent weed may finally be of some use as a source of biogas.

Salvinia can be used to produce biogas, both by itself and as an additive to cowdung.

The Salvinia cowdung mixture produced significantly more biogas per unit feed per unit time, compared to cowdung alone.

When the daily feed in the digester was supplemented with Salvinia the average gas production was 53.3 litres per kg feed per day compared to the gas yield of 29 litres per kg feed per day in digesters of identical design and capacity and operating under identical environmental conditions, but fed with cowdung along.

The gas obtained from the Salvinia cowdung mixture on the modified digester is double the yield from cowdung along on the standard KVIC digester, and a shade higher than the yield from alone on the standard KVIC digester, and a shade higher than the yield from mixed feed on a KVIC digester.

Based on the cumulative experience a new digester costs 15 per cent more than the existing design digester but helps easy feed for the weeds and remove and digested mass.

In the case of existing KVIC digesters, 5-10 kg of the weed can be used with 10-20 kg of cowdung as daily feed, and the scum removed once every two months.

BIO-DESULPHURISATION OF PETROLEUM

Sulphur is the third most abundant element present in petroleum after carbon and hydrogen. The large scale use of petroleum cure in power generation and transportation, results in atmospheric pollution.

Several micro-organisms have been isolated from crude dumping and discharging sites, but their degradation capability was found to decline on continuous exposure to model organic compound debenzothiophene stable mutant from culture. No cardioides sp PKSP for effectively removing sulphur from crude petroleum has been isolated.

Cells of PKSP 12 were pregrown in an inexpensive organic medium, harvested, washed and used for biocatalytic activity as a sell suspension in phosphate buffer. The micro-organism has shown promising results without reducing the calorific values of crude.

RECYCLING DAIRY EFFLUENTS FOR BIOGAS PRODUCTION

Dairy effluents consist of whey, buttermilk and milk plant washings. It is rich in lactose sugar, protein, fasts and some minerals. By virtue of its nutritional composition, it can be recycled for generating biogas.

Experiments show that dairy effluent produce about 40 per cent more biogas, and was found to be a diligent to mix cowdung. This can augment biogas production and save water.

MICROBE FARM TO TREAT EFFLUENTS

Deep shaft, an effluent treatment process, is a spin-off from earlier work using fermentation to produce single cell protein as an animal feed.

That its fermentation technique was not so different to what happened in a sewage works led ICI to scale-up the process and tailor it to the needs of effluent treatment. The ability to make more oxygen available to the effluent digesting microbes is the key to the Deep shaft process. The heart of the methods is a deep cylindrical shaft, measuring typically from half of a metre to eight metres across sunk between 40 and 150 metres vertically into the ground.

The shaft itself, which is lined either with steel or concrete, is divided into two sections, an inner section called the downcomer and an outer section, the riser. The untreated effluent enters the downcomer where injected air forces the mixture of effluent and activated microbial sludge to travel downwards in the downcomer and upwards in the riser in a continuous process of churning and circulating.

The long and convoluted path that the air bubbles must take keeps them in contact with the biomass for several minutes compared to only a few seconds in a conventional process. And because of the hydrostatic pressure in the column of liquid, the solubility of oxygen at the bottom of the shaft can be 10 times that at the surface. The efficiency of oxygenation can reach 80 per cent compared with only around 15 per cent in more conventional processes.

CANNIBAL TADPOLES TO ELIMINATE OTHER SALAMANDERS

There are no lengths to which creatures will not go in the furtherance of their own genes. So, much so, that in the presence of non-kin (or even distant cousins), tadpoles of an amphibian called the Arizona tiger salamander (Ambystoma tigrinum nebulosum), turn from peaceful browsers of pondlife to large, ravening cannibals that consume as many other salamanders as possible.

This tendency is finely tuned to the degree of relationship between salamanders in a pond or aquarium. Should a female lay her eggs in pond unsullied by other salamanders, none of the progeny will, in general, turn against its fellows — unless the population becomes very dense. In those cases where one of the progeny becomes a cannibal. It will do so relatively late in larval life, when it will pose less of a threat to its fellows.

But if two unrelated broods share a pond, the race is on for one of them to produce a cannibal as soon as possible to consume the tadpoles of the other, allowing only its own kin to colonise the pond. In fact, the researchers show that the more different broods share a pond, the sooner cannibalism will develop in one of them.

Nobody would suggest that salamanders are consciously able to discriminate between siblings, cousins and strangers, particularly as the transformation from a normal salamander into a cannibal needs physical remodelling: cannibal mouths and jaws are modified to allow them to consume salamanders of their own size. Given that the transformation involves more than a quick change of costume, why is the cannibal development so finely tuned to degree of relationship?

Cannibalism develops in response to substances released into the water that are detected by small. In the past two years or so, experiments with mice have shown that animals can judge relatedness depending on the degree to which they share genes that contain instructions to build highly variable proteins in the immune system. These variations are, of course, inherited, and breakdown products of the proteins presumably appear in urine, giving the animals a family smell.

TECHNOLOGY TO RECLAIM WASTE LUBRICANTS

Waste lubricating oil can be reclaimed up to 80 per cent without polluting the environment through a technology developed. This technology converts the 20 per cent residue into low grade greases and briquette fuel.

The high vacuum process eliminates the acid treatment of the conventional method. The acid sludge in conventional method causes pollution and corrosion, and is difficult to be disposed off. With the new process the reclaimed oil is as good as the new oil. It is ideal for automotive lubrication and petroleum based oils containing additives one can reclaim oil of required viscosity, which is not possible with the conventional methods.

Economics of the new process include immediate return on investment, low running costs and energy savings. Besides, capital investment is also eligible for full depreciation as the process comes under the scheme of energy conservation.

SULPHUR-EATING BACTERIA

Industrial contamination stretching back a century has started to be reversed by biotechnology. In one of the world's largest electrolytic zinc plants is operated, groundwater below the plant has become contaminated with heavy metals and sulphate over the years.

Scientists have now developed new technologies that have made possible the development of a geohydrological management system made up of pump wells and a biotechnological treatment system for the water produced.

The treatment system uses sulphur reducing bacteria (SRB) to "eat" the sulphate, releasing heavy metals in the process. SRBs are well known in the oil industry as highly undesirable organisms that produce deadly hydrogen sulphide.

DISPOSAL OF ARSENIC WASTE

A novel technology to capture and safely dispose of toxic arsenic waste, with possible application to other heavy metals waste, has been developed.

The process produces a highly insoluble solid to contain the arsenic, which then can be used as landfill.

One popular process was to slurry the arsenic oxide with lime to produce an insoluble calcium arsenite. Unfortunately, this slurry had proved unstable in the long term, breaking down and releasing soluble arsenic into tailing ponds.

Another was to combine the arsenic oxide waste with iron hydroxide, producing a far more durable slurry which, under a wide range of conditions, restricted arsenic movement out of the tailing dams. However, this methods relies on tailing dams being water saturated. If they dry out, wind can blow the arsenic dust about, and if they flood, waterways are contaminated.

According to Dr. White, synthetic arsenic minerals can be produced in the laboratory, two of them promising immobilisation. They are johnbaumite, a rare mineral and an arsenic analogue of merrillite commonly contained in stony meteorites. Both, can be prepared by reaction of the arsenic waste with lime at 500°C. Unlike the calcium arsenite slurries, they are highly insoluble—one hundred to one thousand times less soluble than the tailing dam slurries.

While an extremely durable waste form, a further level of security is achieved by mixing johnbaumite and merrillite with cement to achieve a solid that is suitable for disposal in remote landfills.

LIQUID FUELS FROM WASTE

Agricultural wastes have the potential to be come a major source of liquid fuels and cellulose fibres.

Most of what farmers grow is wasted. Cereal straws contain about 45 per cent cellulose, 35 per cent hemicellulose and 10 per cent lignin. Soft, straws such as rapestraw contain even more hemicellulose, and sugar beet and citrus pulp are mostly pectins.

Cellulose, however, is difficult to convert to its constituent sugars because the fibres are designed by nature to be non-degradable, being semicrystalline and waterproofed by lignin glued in place by hemicelluloses. Hemicelluloses and pectins are easy to hydrolyse but have been ignored because yeasts cannot ferment the resulting sugars.

Efforts to degrade cellulose to sugars seem to us misplaced, since it is a useful natural fibre. Chemical paper pulping of wood hydrolyses the hemicellulose glue and makes the lignin water soluble so that cellulose fibres remain after simple water washing. But the wash contains over half of the dry weight that is uselessly incinerated at considerable fuel cost during chemicals recovery.

However, alternative paper pulping processes are now at hand, using high pressure steam to hydrolyse the hemicellulose glue, which is washed away. The lignin can then be extracted with organic solvents such as ethanol. One process used hot aqueous ethanol to combine both steps — the ethanol is recovered by distillation, leaving soluble hemicellulose and undegraded lignin. The latter could make useful fuel for the plant and the former could be converted to ethanol for addition to petrol as an octane enhancer.

Since yeasts cannot efficiently utilise the major sugars derived from hemicellulose or pectins. Research engineer an organism that can do so.

A thermophilic bacillus utilises hectics etc. anaerobically very rapidly at 70°C, probably because it natural home is high-temperature compost heaps. It did not make much ethanol, but a mutant by a combination of genetic and physiological manipulation makes ethanol from pentoses in yields approaching that of yeasts on starchbased sugars, but much more rapidly.

Since non-growing cells give maximum ethanol yields, a novel continuous two-stage 'closed system' maximises ethanol production from a wide range of mixed sugars. Hemicellulose sugars are fed to nongrowing cells at 70°C retained in an anaerobic reactor by membrane filtration. Most of the ethanol is removed in the vapour by nitrogen sparging or mild vacuum — this allows use of high sugar feeds without ethanol toxicity. Ethanol is stripped from the filtrate by distillation or membrane separation and the stillage is passed to an aerobic fermenter in which the unused sugars and by-products are converted to flesh thermopile cells. These are then concentrated by membrane filtration and passed to the catalytic anaerobic reactor to compensate for the death rate of the cells therein.

Safer Method of Removing Asbestos

An innovative and relatively simple method of asbestos removal without endangering operators and by-standers has been developed.

Asbestos has produced a major industry in its removal to prevent the carcinogenic diseases of asbestos and mesothelioma.

Extracting asbestos from buildings need not be expensive, disruptive or dangerous to people working with or near the operation. Using special water sprays to collect asbestos dust at its source by forming a liquid barrier, it eliminates the need for space suit type protective clothing or for special restrictions on the building being decontaminated.

It removes asbestos from pipes and other equipment by placing the item being treated in a rectangular PVC case with nozzles directed to surround the pipe with water spray. Using a special process, the sprays create a fine aerosol which sticks to the asbestos fibre.

As the dust is trapped by the spray, air pumped into the case directs the waste water through a drain to a treatment plant where the asbestos slurry is removed as the water treated for reuse. The sections of asbestos surrounding the pipe also are saturated with water to prevent any emission of fibres. The asbestos then is pulled free by the operators wearing heavy duty gloves, and disposed of in double bags of 200 micron PVC to further ensure safety.

The system provides equipment to suit all pipe configurations whether horizontal, vertical, long radius bends, brackets and stauchions.

ELECTRICITY FROM MUNICIPAL WASTE

Production of large quantities of waste materials such as municipal waste had become pollution hazard. Fortunately the new technologies help us to combat such hazards by way of innovative methods to usefully dispose off the solid wastes.

Apart from safeguarding environment, such methods result in recovery of useful metals and production of electricity in the process. Even after recyclables are removed in the process, the refuse is not an ideal fuel. It still varies widely in size, composition and moisture content. But it can be transformed into a relatively homogeneous mixture of uniform size, heating value and moisture content, if suitable equipment is used.

For full RDF (refuse derived fuel) firing, municipal solid waste must be prepared to a uniform size, generally 127 mm or less in any dimension. It is also necessary to remove as much of the non-combustible material and abrasive fines as commonly used for this purpose.

New Venture

One such resource recovery venture is at Ohio, Hawaii, with a population of 8,34,000 people. Also five million people visit this island every year, if land filling is resorted to dispose the over seven lakh tonnes of solid waste produced annually. The land would be consumed in a few years apart from causing environmental and health problems.

In the Honolulu resource recovery venture, the daily volume of processed solid waste is burnt in steam generators, (equipped with ABB dry scrubbers and electrostatic precipitators) to produce electricity. With over five lakh tonnes of waste processed annually, the facility generates sufficient electricity to serve the needs of about 40,000 homes.

Further, the generation of over 300 million kilowatt hours of electricity each year at the plant has allowed the island to reduce its annual consumption of imported oil by nearly over 570,000 barrels valued about $ 12 millions.

Process

The facility (capacity 1,960 tonnes per fay) uses ABB resource recovery systems proprietary waste processing technology. Municipal solid waste is processed into a refuses derived fuel by removing the magnetic ferrous metal and non-combustible materials and by shredding the remaining combustible materials to a size suitable for burning. The cost of this facility is about $ 150 millions and the service cost is $ 15 millions a year.

The non-combustible processing residue as well as the ash resulting from the burning process are being used as landfill. Steam produced fuel (RDF) is used to generate up to 57 MW of electricity. The fuel value of the waste is about 68 gallons of oil per tonne of solid waste.

The plant is located in a 24 acre site in Hawaii. Municipal solid waste is transported to a refuse storage and feeding area (maximum capacity 5,500 tonnes). The waste is fed into conveyors which carry them to primary shredders. The magnets separate the ferrous material from the waste. These ferrous material are removed to the recycling plants. The primary separation units separate the residue from the waste and the secondary separation units further separate the non-combustible residues which are used for landfilling purposes. The shredded waste is taken to a secondary shredder which processes the waste further.

The process fuel is taken to reduce derived fuel (RDF) storage (up to 2,540 tonnes) through conveyors and then transported to steam generators. The processed fuel is burnt in the boilers to produce steam (up to 1,13,400 kg per hour) which drives the turbo generator to produce electricity.

The electrical energy is distributed to the consumers from a switchyard. An air quality control equipment removes the dust from the flue gas before the gas is emptied through the stack.

The system consists of two identical processing lines, with a common line equipment that serves both processing lines and the necessary electrical controls to start, stop, monitor and control all the components. Each line processes at an average 91 tonnes of waste per hour producing a nominal RDF particle size of 90 per cent less than 10 × 10 mm. Through-put rates will fluctuate depending on the type of refuse, time of year, speed rate of infeed conveyors and condition of various equipment such as shredder hammers, shredders and magnets.

Adoption in India

The facility described above may be an ideal one to beneficially dispose off the solid waste. Of solid waste is being collected and disposed off by mostly filling low-lying areas. The quantity is bound to increase considering that the average solid waste (excluding human excrete) generated in the cities work out to 250 grams per head per day and the population is increasing at three per cent per year.

The cost of generation of electricity from the facility in the U.S. is about Rs. 8 crores per Megawatt. This as such is a bit higher but will really amount to a much lesser rate reckoning the cost of collection, transport and disposal of the wastes and prevention of menace due to breeding of flies. The probable cost as per Indian conditions may come to Rs. 4.5 crores per MW. Since this is a new source of energy, and development of this technique will result in getting "Wealth from waste". It requires serious consideration. (S. Ramaswamy)

POLLUTION CONTROL DEVICE FOR CARS

Motor car manufacturers will soon be able to buy indigenous catalytic converters to control pollution from vehicle exhaust.

A "honeycomb" structure for housing the catalyst will break down the noxious exhaust gases into harmless products.

The honeycomb is a vital component of the converter and only a few companies in to world were making it. The honeycomb structure having 200 cells a square inch has been developed.

INDIGENOUS ELECTRIC VEHICLES

The advanced pure electric/electronic high performance road models up to two tonnes payload capacity and industrial models, tractors and platform trucks, from 5 to 20 tonnes capacity. Their life is about 20 years.

The electric tractor with rheostatic speed control suffers from the loss due to Joules 1^2H heating and the rapid war of resistance contactors. Hence their life is limited.

The electric/electronic vehicle (EV) is operating in several States and the technology is now 93 per cent indigenous.

There three main types of EVs.

The converted internal combustion road vehicle, is built on a Matador chassis retaining the IC gear transmission system for reducing cost. This hybrid type suffers from the incompatibility of the SCR chopper with the mechanical transmission resulting in frequent break downs.

These electric/electronic vehicles have lives three times longer than the equivalent Internal Combustion types. Also operating and maintenance costs are much lower. Infrastructure facilities for servicing and repairing exists in several States.

The high performance EV types in the country have the following specifications.

Range : The latest road models have a range of 70 kms per hour, which can be readily extended up to 140 kms by fitting a quick change second battery system. Range is not a criterion for industrial models for as these are designed for opportunity charging between duty cycles.

Speed : The latest road models have a speed of 50 kmph. Industrial models have the advantage of being much smaller for factory and workshop operation than the equivalent IC vehicles.

Gradient capability : The DC along carcass traction motor has a higher gradient capability than most other vehicles. Their starting acceleration is not equalled by any other type.

The batteries can be recharged. The lead-acid batteries have a life of four years. (Indra Chatterji)

CLEAN PLASTIC WELDING

The resistive implant tape welding (RITW) process, developed involves heating an implant element at the joints of sheet materials to form high quality welds. The secret is a polymer tape that contains a copper conducting element and a thermoplastic that melts while welding. The tape grips the material to be welded and exerts pressure. It also provided additional welding material to avoid thin coating problems. Welds up to 100 m can be made in about three minutes.

Trials carried out on a range of polymeric materials have not only proved welds to be to engineering levels of strength, but also to have guaranteed repeatability. The method means more reliable joints in safety ability critical areas.

V

CLEANER AND POLLUTION-FRIENDLY TECHNOLOGIES

Search for Cleaner Autofuels

The automobile industry prefer to use petrol and diesel. Pushed further to reduce emissions, they would opt for cleaner gasoline. (Lead in petrol is discontinued in many countries.) Gasoline can be reformulated to meet stricter standards of emissions to reduce toxic emissions and volatile organic compounds by 15 per cent. And if a minimum of two per cent oxygen is introduced into petrol, emission of carbon monoxide can be lowered. But then consumers will have to pay more for such fuels as it involves a costlier refining process.

In contrast to an alternative fuel which would require expensive vehicle modification, reformulated gasoline could be used in all conventional automobiles.

Alternatives

Natural gas — oil's little brother — is abundant. And recent reserve additions for gas have exceeded those for oil. What's really holding up the greater use of gas is the high cost of gathering, and transportation especially over the oceans. It becomes economic when transported in pipes in large quantities.

Compressed natural gas (CNG) for use directly as motor fuel requires costly (over U.S. $ 1,000) modifications in the engine. Also there is difficulty in storing enough gas to give a reasonable range. CNG under pressures as high as 250 bars (25,000 kilo Newtons per square metre) is one solution that is being adopted. But still the compressed gas occupies four times the space of equivalent gasoline, besides the added hazards.

Problems with Methanol

The likely approach is to convert methane into liquid methanol, which could be used in auto engines without major modifications. But liquefying natural gas isn't simple. Several technologies are available but none can match the prices of gasoline. In addition pure methanol gives starting trouble at temperatures below 10^{0}C, is as expensive as gasoline and provides only 70 per cent of the mileage. Petrol may also have to be added to give stability to be volatile alcohol. Recent research particularly at Volvo's assume a mixture that has 15 per cent petrol.

Liquefaction of Coal

Technology for converting coal into methanol through liquefaction has been there for some time. As coal is a hydrogen-deficient hydrocarbon (its hydrogen: carbon ratio is near 0.8 compared whereas for petro the ratio is near two), any process to convert coal to alternative fuels must add hydrogen (either directly or in the form of water). That's why the process reuses immense quantities of fresh water.

The following methods are used for coal liquefaction :

Pyrolysis (heated in a stream of hydrogen).

Solvent extraction (coal hydrocarbons are dissolved and hydrogen added).

Catalytic liquefaction (hydrogenation in the presence of a catalyst).

Indirect liquefaction (carbon monoxide and hydrogen are combined in the presence of a catalyst).

Work is on to make liquefaction commercially viable and environmentally acceptable — mining, handling, and burning coal can pollute. Also, many technical problems remain to be solved. Probably they would be good enough for boilers in the first instance. Refining synthetic crude to various exacting specifications for automobiles would be a complex undertaking.

Oil shale (biggest deposit in the U.S.), occurs mostly as stones (similar to limestone). It is necessary to heat the solid to over 500^0 to release the liquid shale oil. With the best deposits it is possible to recover 25 to 30 gallons of oil per ton of rock. Therefore, the amount of material to be handled would be enormous for large-scale operations.

The crude extracted from shale has higher viscosity than petroleum crudes. It calls for costly refining for use in automobiles.

Methanol has a few environmental advantages. Notably, its widespread use could reduce relative hydrocarbon emission anywhere from 35 to 85 per cent, decreasing smog formation sizably. But it fails to reduce carbon monoxide emission and might even increase emission of carcinogenic formaldehyde by five to ten times compared to gasoline.

Methanol is also corrosive and so automobiles must be modified — increasing prices by about US $ 300.

Ethanol Goes Out

Ethanol, is an alcohol fuel made from corn, molasses and other organic substances. When added to gasoline, it reduces carbon monoxide emissions to a minimum of 3.1 per cent by increasing the oxygen content. The widespread use of ethanol is in Brazil, where over a million vehicles run on "gasohol" — a mixture of 78 per cent petrol and 22 per cent ethanol. Another million vehicles run on pure ethanol. All these are estimated to have reduced carbon monoxide emissions by 20 per cent, besides marginal reduction in nitrogen oxides, while reducing mileage by only two per cent.

However, like methanol, ethanol is more expensive than gasoline. It is also far from environmentally sound. Being highly volatile it evaporates more quickly than gasoline. Its widespread use can a raciable increase volatile organic emissions step up urban ozone levels and lead to smog. Its use could increase other pollutants such as carcinogenic aldehydes. And engines have to be adapted to the increased corrosivity, by coating the insides of fuel tanks with tin, the fuel lines, with copper and the carburettor with zinc.

The search for cleaner fuels should continue. (Mohinder Singh)

ENVIRONMENT-FRIENDLY CHEMICAL PLANT

Chemical industries the world over are known for their environmental pollution.

Memories of Minimata

In the Seventies the mercury cell process was commonly used for production of caustic soda. (In fact, today there are around 30 caustic soda plants in India using this process.) Now, the very mention of mercury brings to the lay mind the horrific Minimata Bay eco-disaster in Japan in the Sixties which led to many persons in the area getting paralysed due to eating mercury-contaminated fish. But, the fact is that the Minimata Bay disaster was due to mercury bearing effluents from PVC (Polyvinyl Chloride) plants being let out into the Bay and the mercury consequently accumulating in the fish tissue.

In caustic soda manufacture, there is actually no such thing as a mercury-bearing process effluent since the mercury is totally recycled within the process loop itself. However, some mercury about two to 10 parts per million (ppm) — can surface in the floor washings through leakages from the cell house and concentrator. The allowable limit of mercury in the discharge into the sea was 0.01 ppm.

The first step towards developing a treatment process was to install an Atomic Absorption Spectrophotometer to detect mercury at ppm and lower levels.

Towards a Clean Plant

A simple, batch-type process was evolved. The first step comprises removal of the suspended matter followed by clarification and pH adjustment. Then comes a chemical treatment with a special formulation which converts all the different soluble formulation which converts all the different soluble norms of mercury in the washings into cinnabar (Mercuric sulphide) which is insoluble and can be separated.

Further treatment of the washings consists of reaching with activated carbon and then cleaning up in an ion exchange column. This ion exchange process was developed in-house and was found less costly than those available abroad.

The final effluent contains less than 0.005 ppm of mercury.

The mercury content in the treated washings was far below allowable limits. To totally recycle the washings from the caustic soda plant instead of letting out the treated washings into the sea, the wash water was treated further to remove the sulphate and then used as part of the process water to make hydrochloric acid which is used in the phosphoric acid plant.

Sea Discharge Safe

The only liquid waste that now flows out into the sea is a solution of calcium chloride (100 gms per litre) from the phosphoric acid plant — 600 kilolitres per day discharged through a pipeline that extends 800 metres into the sea. Incidentally, seawater itself contains about one gm per litre of calcium chloride.

Radioisotope tracer technique studies conducted have ascertained that within three metres of the discharge wind of the pipeline, the effluent is fully diluted several thousandfold. The plankton count showed an increase in the plankton count and healthy state of marine life.

The stack gases, when analysed, showed a mercury level of 10 milligrammes per cubic metre, whereas the permissible limit was 300 microgrammes per cubic metre. A special absorbent, a carbonaceous material is chemically treated to entrap the mercury in the stack gas. With this material, it was possible to reduce the mercury level to less than 10 microgrammes per cubic metre.

Solid Wastes Problem

The one remaining problem was solid waste. This waste contains calcium carbonate, magnesium hydroxide, barium sulphate metal hydroxides, earthy matter and about 100 ppm of mercury. As of now the solid waste, is being dumped in selected impervious-bottomed lagoons within the factory campus.

Management Initiative

Improving the surrounding environs through an aggressive tree planting drive. Among the varieties planted are acacias, subabul rain trees, gulmohars, casuarings, jumun, cassias and many others. Using proper fencing and regulating entry of cattle, the survival rate has been as high as 70 per cent. Of Aperimeter wall around the campus keeps the wood poachers away.

The consequence has been a resurgence of wildlife us, especially a wide variety of birds and small creatures such as rabbits and mongoose. (N.N. Sachitanand)

CONTROLLING MERCURY POLLUTION IN CHLOR-ALKALI INDUSTRY

The natural sources of mercury pollution are volatilisation and leaching of dimethyl mercury from geological structures and formations. The man-made polluting sources are discharges from chlor-alkali plants and chemical industries, paper and pulp plants and pesticide companies. Metallic mercury is less toxic to living organisms but in its organic form (say methyl mercury) the toxicity is high.

The level of mercury concentration in air in Indian industrial atmosphere (processing mercury mg/m^3 against a threshold limit of 0.1). In India more than 200 tonnes of mercury is consumed by chlor-alkali industries every year. Even the contamination due to earlier discharge is so high that it would take many years before being assimilated by nature.

Considering the importance of controlling mercury pollution in chlor-alkali industries, the quantity of mercury bearing effluent should not exceed 10 m^3/tonne of caustic soda produced. It has also been stipulated that the quantity is to be further reduced to two cubic metres per tonne of caustic soda produced with a mercury content of 0.005 mg/1 of effluents discharged. The time limit it yet to be fixed. The standards stipulated are so stringent that effective control measures are to be adopted.

In India caustic soda is mainly produced by mercury cell process where a concentrated solution of sodium chlorides and a moving stream of mercury cathodes. The sodium liberated forms an amalgam with mercury while chlorine is evolved as gas at the anodes. The sodium amalgam flows into a secondary compartment where it is reacted with water to produce caustic soda and hydrogen gas. Theoretically, mercury should not be lost in process but there is a steady loss of mercury.

A major portion of the mercury entrained in the washing can be recovered and used physically by providing traps in the drains. However such measures do not completely stop the entry of mercury needing further treatment.

Different processes are available for recovery of mercury from effluent streams; ion exchange precipitation with sodium sulphide, or iron sulphide or hydrazine.

Choice of Process

Ion exchange requires resins of desired type, and capacity. Earlier resins were imported but now resins are available indigenously, though at higher prices. Plants are normally designed for a maximum flow rate of 25 m^3 of effluents per hour. For a plant of 25 m^3/hr capacity 1500 litres of resin would be required. Normally resin is to be replaced every two years.

Resin absorption pretreatments of the effluents like chlorination for conversion of mercury into anionic form and then absorption of the excess chlorine on activated in traces, will damage the resin.

As for the hydrazine process, it is not attractive due to high capital investment and complete removal of mercury is impossible.

With precipitating mercury with sodium sulphide and iron sulphide, the main problem is that the precipitate formed is highly colloidal and filtration indifficult.

Process

The treatment involved the use of sodium sulphide too convert mercury compounds into insoluble mercuric sulphide. The precipitate is removed by filtration and the filtrate is found to have a mercury content of 003 mg/1. Though the reactions are carried out batchwise the process can be carried out almost continuously.

The mercuric sulphide precipitate is removed and converted by sodium hypochlorite into a brine soluble complex Na_2HgCl_4. This can be recycled back to the cell along with the feed brine where it will be reduced to metallic mercury. Another way of recovering mercury from this precipitate is to take it out physically from the filters that mercury is evaporated and condensed. Thus pure mercury can be separated.

The plant can be modified to any size depending upon the requirement. (U. Chandrahasan)

Pollution-Friendly Natural Gas-Powered Car

A natural gas-powered passenger car has been developed in Japan.

The car features a specially designed high roof which houses four natural gas cylinders. They are made of an aluminium alloy covered by fibre reinforced plastic (FRP). For reinforcement, the aluminium ally cylinders are wrapped in glass fibres that absorb epoxy resin, which are then hardened by thermosetting. The result is lightweight cylinders with a total volume capacity of 24.8 cubic metres.

Featuring an exclusive natural gas engine with enhanced performance, thanks to a high compression ratio, this automobile is expected to travel approximately 330 km on four cylinders of natural gas.

Natural gas engines have a driving performance which is almost compatible with gasoline and diesel engines. In addition, natural gas is composed primarily of methane which contains only a small amount of carbon. Therefore, the amount of carbon dioxide produced is minute. Because of this environmentally-friendly aspect, natural gas is receiving attention as a clean alternative fuel for oil.

OZONE-FRIENDLY FIRE-FIGHTING

A new fire-extinguishing gas mixture made from common atmospheric gases could crust the ozone destroying halon 1301 industrial fire-fighting systems.

Halon 1301, otherwise known as trifluorobromomethane, is used in most fixed full-flooding fire-extinguishing systems which protect computer suites, engine rooms, oil rigs and aircraft by flooding rooms with the gas, straving the fire of oxygen. But halon 1301 has a great potential for destroying ozone in the Earth's atmosphere: ten times that of CFE-11, the reference point for calculating ozone depletion potentials. It is estimated to survive in the atmosphere for 110 years.

The new gas mixture called Inergen is a spin off from NASA space programme and is a mixture of nitrogen, argon and carbon dioxide. When it is mixed with air to concentrations of up to 40 per cent, it reduces the level of oxygen from the usual 21 per cent to between 13 and 15 per cent.

Argon is included to equalise the density of the mixture with that of air. This ensure between mixing than with nitrogen alone, when would tend to rise to the ceiling. The carbon dioxide is present to stimulate the respiration of anyone trapped in the area where the gas is released. The respiratory system is highly sensitive to any changes in concentration of carbon dioxide, responding to increases through deeper, more vigorous breathing.

Inergen is inexpensive, being a simple mixture of commerical gases. Converting an existing fire-protection system to Inergen can use existing halon pipework but the release nozzles and storage tanks will have to be replaced.

ENVIRONMENT-FRIENDLY TEXTILES

Standards had already been drafted or were under preparation in some EC countries which specified certain maximum limits for pesticide residues on fabrics made from agriculturally cultivated fibres. Others standards prescribed limits for dye residues on fabrics and even forbade the use of certainoze dyes which were supposed to be carcinogenic. Fabrics coloured with benzidine dyes may also soon be banned.

Standards were not yet statutory but the 'eco' label for textiles is getting increasingly popular among customers in Europe and many eventually become mandatory.

To fix these dyes salts are used and if such salts like chlorides are high in quantity, the dyed fabric could do more harm to the skin than low impact chemical dyes. The problem is that fabric processors do not give adequate attention to using just the right amount of dye conventionally used can suffice without sacrificing the colour range.

The products of the handloom fabric makers may be more eco-friendly than that of several big mills. The big mills are tempted to employ cheaper and less safe dyes in order to reduce the cost of processing large product volumes.

BATTERY CARS

A battery powered car is extremely quiet, does not emit any exhaust gases, and the power to recharge the battery comes out of the next electrical socket.

This last characteristic is a reason why the electric-powered car should not be given unrestricted backing, for the socket gets its electricity from a power plant, which mostly is run by gas, oil, or nuclear power. What this means is that the environmental impact of a battery-powered car is shifted to the site of the power plant.

The use of such cars in terms of fuel consumption and exhaust emissions will depend decisively on how the cars are used and what power plants the electricity is supplied from.

In city traffic, electrical cars clearly come out ahead in such consideration, with their energy consumption 20 per cent below that of comparable petrol powered cars.

One reason for this is that when waiting at traffic lights, a petrol powered car still burns up fuel while the engine is idling, whereas the electrical car uses up no energy. But this fuel-efficiency advantage of electric cars is complete loss when comparing the petrol-versus-battery performance of cars out for longer distances and higher speeds on the open road.

In terms of exhaust emissions, the electric cars come out better than petrol driven cars provided that the electricity generation comes, as in Germany's system, one-third from lignite, one-third from nuclear power, one-quarter from hard coal, and the rest from other energy sources such as hydroelectric dams.

Electrical cars would show an even greater advantage over petrol-driven cars. Short distance city traffic would handled by an electrical motor, which would be replaced with a petrol powered engine for long-distance driving.

Batteries must be the exclusive source for the exhaustfree cars performance, including all accessory equipment.

In the summer heat, it remains to be seen how electrical cars will be able to provide the hugely energy consuming air conditioning.

NATURAL GAS POWERED CAR

A retrofitted compressed natural gas (CNG) engine in a prototype CNG car with on-board gas storage was tested.

This demonstration has proved the adequacy of the gas carburettor.

The merit of the CNG engine is that carbon monoxide and hydrocarbon emissions are much lower compared to gasoline operation. The gas induction unit consists of pressure regulators for supplying gas in appropriate quantities in predetermined air-fuel ratios. The engine can be operated on their CNG or gasoline as switching is instantaneous.

The specific energy consumption of the machine remains comparable. Marginal loss of power is observed near maximum speed. As an off-shoot of this project a truck mounted cylinder cascade for transporting CNG, a filling-station and dispending facility has been developed.

'Clean' Car goes a long way

International Automotive Design (IAD), has produced a prototype vehicle that combines Nineties car styling with the environmental benefits of an electric car. The up-market design looks certain to be one of the most desirable cars to drive IAD to solve its dangerous smog problem through a most unusual route.

By 1998, two per cent of all cars sold in the metropolitan area will have to be pure electric, or zero-emissions, vehicles.

Hybrid Solution

A novel hybrid vehicle, incorporating a small additional petrol engine to keep the battery charged up on longer journeys.

The electric car still needs an eight-hour overnight charge to keep the batteries topped up, the conventional engine is a backup if the batteries go flat. In fact, the car includes a 32-litre tank, sufficient to take the driver a longway along the highways.

Auxiliary Power Unit

It offers in its all-electric mode is said to be sufficient for a minimum of 90 per cent of all car trips in urban areas. However, one of the design briefs given to IAD was to produce a vehicle that could operate for more than 240 km. Hence the inclusion of an auxiliary petrol-driven power unit a low-emission internal combustion engine equipped with a pre-heated catalytic converter. The goal has been to produce a zero-emission vehicle which overcomes the range limitations traditionally associated with electric cars. Even when the gasoline engine is running, emissions stay below the very strigent ultra-low emission standards.

Other design features have been included to help minimise power consumption. These include aerodynamic body styling, especially developed low roll-resistant typers, a two-speed automatic transmission and an advanced electronic power train management system. IAD even specified a heat repellent type of glass to help reduce the power demands of the air-conditioning system. Air-conditioning is achieved through a solar powered vent.

Sealed Battery System

The key to an electric car is obviously the battery system. IAD scoured the world for suitable batteries and eventually decided on an array of sealed recombination-type lead acid cells. The main advantage of a sealed system is that it is safe and needs little maintenance. IAD claims the motorist will never have to see the battery pack, even if the car rolls over on its back. Although this type of battery has been around since the beginning of the century. It is still the most reliable.

However, the model is engineered so that it can use the latest sodium sulphur and nickel cadmium batteries. The batteries will probably three or four years, with costly replacement. Despite high cost, IAD believes the overall maintenance and running costs will less than for an equivalent petrol-driven car. The savings would be even greater if they ever goes on sale in countries where petrol are higher.

The propulsion system is a combination electric motor and the auxiliary petrol engine. The electric motor drives the front wheels directly through the purpose built two speed automatic gearbox. The petrol engine drives the front wheels through the same gearbox. Its maximum performance can be up to 25 KW if needed, but it only comes into play if the power train management system registers that the journey is more than 65 km. The power train system can give sufficient acceleration for joining freeways and sustained hill climbing. A dashboard display allows the driver to select ''battery only'' or ''assisted motoring''.

A technique known as regenerative braking turns the motor into a generator when the car is slowing down or travelling downhill. This reduces brake wear and has the added bonus of recovering energy for the battery.

The car can accelerate from 0-80 km/h in 17 seconds, and has a top speed of 120 km/h, not much lower than many small cars available to-day.

DEVICE CLEANS UP POLLUTING

Up to 40 per cent less exhaust pollution and the same reduction in petrol consumption may seem like an unattainable dream in these days of car-polluting cities but that is the claim of a device invented.

The device is particularly effective on older motor vehicles which have not been modified to reduce polluting exhaust gears. When these vehicles are stationary and idling up to 40 per cent of fuel passing through their engines is discharged unburnt as pollution.

The device is an electron stimulator which aims to complete combustion of petrol by reducing fuel globules passing from the carburettor to the manifold to molecular size. The atomisation of fuel enables the fuel molecules to bond almost completely with oxygen atoms and thereby burn almost totally.

The electron stimulator consists of an aluminium moulding containing two or more neodymium iron boron super strong magnets.

Recently completed tests indicated the stimulators reduced the number of large particles and increased the number of small particles. This change was large enough to affect fuel combustion.

The trail showed that the stimulator atomised fuel down to at least three microns, the smallest measure possible with the method used. Field trails showed that the stimulator brought the average hydrocarbon emission to 139 parts per million from 520 parts per million.

Another test on racing engine showed that with the fitted stimulator the engine emitted hydrocarbons at a rate of five parts per million at 6500 rpm. Without the stimulator it discharged hydrocarbons at a rate of 515 parts per million.

Reducing Emissions from Industrial Gas Turbine Clean Burn

Environmental protection techniques of steam/water injection develop a new type of combustion process for gas turbine. The solution developed aimed at reducing emissions of NO (oxides of introgen) and CO (carbon monoxide) simultaneously, using an engine modification known as the dry low emissions (DLE) system.

Operational Flexibility

The system provides the operator with full operational flexibility — unlike the earlier water or steam injection systems offered to reduce NO_x emissions, which cut operating flexibility. In addition, the company recognises that engine operators in natural gas transmission do not want water injection at all.

The system has been launched to achieve the lowest possible emissions when operating on natural gas, without using any other "working fluids", while maintaining the compact arrangement that distinguishes the aerospace-derived gas turbine.

With conventional gas turbine combustion systems, the weakening of the air-fuel mixture is limited to a ratio of about 20:1 in the primary combustion zone. With this technology ratios of up to 60:1 are possible but the problem of "weak extinction" (in which the flame goes out) occurs at a ratio of 40:1. To overcome this, engineers devised a system of "staging" fuel through a series of three zones, providing greater operating flexibility.

Combustion Temperature Control

The alternative of "parallel staging" involves multiple points of fuel injection, and a fuel schedule in which all zones are variable until the weak extinction margin is reached, when the required number of zones are extinguished to enrich the remainder — a process that can be repeated throughout the power range. But it is much more difficult to obtain full operating flexibility by this technique.

A premix lean burn system is adopt to control the combustion temperature without in the tight "window" that occurs around 1,700°C, when both NO_x and CO emissions are at their lowest. Fuel staging provides turndown capability over the required load and ambient temperature range while avoiding flame out, and the series fuel staging system provides a wide stabilivoid chilling of combustion areas and a consequent rapid rise in CO emissions.

Retrofit Module

In single-combustor rig testing, the company claims to have achieved an emission level significantly less than 25 vpprn NO_x, with zero CO, under conditions representing full power engine running.

The revised engine design required to incorporate the new combustion system retains the modular construction in which five modules, each fully interchangeable with factory-overhauled replacements, facilitate maintenance with low downtime.

Changes in design are restricted to the high-pressure module (module 04), to simplify retrofitment, and the new system will be available generally as a retrofit package for existing engines by modular exchange, or as a new power plant, in 1994. The module comprises nine tubular combustors volume-fitted between the existing high-pressure (HP) compressor and turbine sections of the engine.

Dual Fuel Development

Performance in conjunction with a typical power turbine is around 47,000 to 54,000 KW power for mechanical drive or comparable electrical output. With the new package, the engine appears set to become even more popular with today's environment conscious operators.

WASTE HEAT FOR REFRIGERATION

A vapour absorption refrigeration (VAR) system (capacity 50 tonnes) using waste heat emanating from diesel generator exhaust gases as innating from diesel generator exhaust gases as input has been designed and demonstrated.

"Commercial size VAR systems are available to industries where large amount of waste heat is available and simultaneously a demand for refrigeration exists," said Mr. Rudra Dev its executive director.

The need for the conservation of energy has revived the interest in VAR system technology. In chemical industries, waste heat in the form of process steam and flue gases is available. Approximately 50 per cent of it is available as low grade thermal energy temperatures between 100°C and 200°C. This energy could be used as heat input to VAR heat pumps and bottoming cycle power generation systems.

VI

FOLLOWUP-LOOKING FORWARD AND THE ROAD AHEAD

The post-Rio action should consider some of the key issues like Sustainable Development (CSD) without jeopardising the environment.

Environment should not be used as a potent weapon in the North-South political and economic battle. The affluent countries fight hard to keep their high-cost economics going while the environmental scenario demands an entirely new approach to development.

While the North has developed over the least 150 years on cheap and plentiful energy and on reckless exploitation of the earth's natural resources, environment is now being used to slow down development in the South, while the North refuses to give up or alter its ecologically expensive and otherwise wasteful life-style.

Both local and global environment issues should be kept in mind while planning. Global environment issues be addressed in an equitable and democratic manner without jeopardising development.

Poverty and development could not be isolated from environment issues. In implementing Agenda 21 and in setting up of a democratic CSD, the North is not to be allowed to dictate terms to the developing countries.

We should insist to set right the global environmental wrongs, the industrialised nations responsible for damaging to the earth's environment must pay for all incremental costs of steps the developing nations have to undertake. This must be over and above all other on-going development aid.

The environment-conscious voluntary organisations in the North are also increasingly becoming conscious of the stand taken by the developing nations.

CRUCIAL ISSUES FOR FOLLOW UP — AGENDA 21 FOR ACTION

Crucial issues related to phase out ozone destroying chlorofluorocarbons (CFCs) by harmless substitutes.

We envisage a state of zero production and consumption of CFCs worldwide by the year 2010. Funding of transfer of technology related to substituted for CFCs, the question of interim technology, the problem of halons which are essential for fire-fighting and defence uses the issues of delinking trade sanctions and transfer of technology are to be scheduled.

Unless developing countries were given the means to phase out the old EFC based technology, unless the issue of transfer of technology is addressed fairly and squarely, we could not successful.

The countries which have the new ozone friendly technology must make it available on terms that make it possible for developing countries to acquire it through funding from the interim fund created for the purpose.

One of the important issues is the issue of conversion of this interim fund into a permanent fund. Currently contributions are voluntary and it has about $ 50 to 60 millions, but it would have to be raised ten-fold to enable it to meet the incremental costs of the switchover in the developing countries.

The ozone problem has been created by the developed countries and they must pay for the switchover. We should find a way to avoid an interim technology using HCFCs and move on to the third generation technology considered entirely environment-friendly.

We should ensure that funds are made available for research and development work in the developing countries, to get out of the situation which reinforced latter's dependence on technology developed by the industrialised nations.

'GLOBAL OBSERVING SYSTEMS MUST BE STRENGTHENED'

The global observing system must be urgently strengthened worldwide, especially in developing countries, where there are major gaps in the systems. These improvements are needed to narrow the uncertainties in the timing, magnitude and regional patterns of climate change. The establishment of the global climate observing system (GCOS) is part of that required response.

''Environmental crisis'' threatening the future of the planet including *Global warming, the depletion of the ozone layer, deforestation, water and air pollution*: these are considered the most worrisome threats jeopardising life for future generations. To evolve a course of ''sustainable development'', the post cold war world had to come to terms with new geo-political and economic realities and also urgently seek solutions to the ecological challenges confronting the earth.

Undoubtedly, the environment crisis is of crucial concern for the survival of the entire planet. But what was overlooked or glossed over is the very specific micro-level reality that most countries have to contend with — *grinding poverty, famine, the lack of drinking water, societies with high infant mortality and poor life expectancy, a degraded environment and health hazards that make even day-to-day survival a big question mark.* This crisis is what needs immediate attention and equally important is the fact that the *present ecological crisis is directly attributable to the unsustainable consumption and production patterns of the industrialised countries.*

The basic dissension lies in the fact that the South perceives the North countries as being responsible for this state. The U.S. for example with only 5 per cent of the world's population, consumes 25 per cent of the world's energy and emits 22 per cent of all the carbon dioxide.

The action programme covers a wide range of environmental issues including *desertification, deforestation, water, agriculture, and development issues such as a health and education.*

VII
SOME RECENT DEVELOPMENTS ON THE ISSUES DISCUSSED IN THIS BOOK

1. Environmentalism and Globalisation

The recently emerging environmentalism is a much more powerful globalisation and universalisation process than the existing ones like satellite television, revolution in Information and Communication Technologies (ICTs) etc. Integrated Pollution Control of the whole land, water and air do overlap each other and become global. Global policies affect resource — defletion, extravagant consumerism, life styles, population explosion, bio-diversity base of the Earth. Environmental degradation is the consequence of all economic and industrial activities like production, waste-disposal etc. Global warming, ozone depletion, bio-diversity conservation should be guided by scientific methods rather than political interests (See below).

2. Bio-Sphere Preservation

Scientists must work out the right proportions in which plants, forests, animals and the humans should take their place on the Earth to ensure that the planet remains habitable long enough for all of the above. We should protect one bio-reserve base and prevent the loot and plunder of our biological material like medicinal plants, fibres, oil, cancer-curing indigenous herbs, local flora with healing properties. Illegal bio-prospecting by foreign agencies should be stopped. Laws must be enacted as agreed to during the Bio-Diversity Convention in 1992 to protect the genetic resources. "Prior Informed Consent and Material and Information Transfer Agreements" should not be violated or abused. To regulate bio-prospecting, the concerned agencies should have to negotiate, with the concerned Authorities, seek prior 'Informed consent' and share the benefits with the locals.

3. Environment and Development

Priorities should be set focussing on human beings and the eco-system. The existing dichotomy between environment and development must be *erased*. The key to building synergies between development and environment lies in recognising the fact that proper development helps environmental protection and vice-versa. Efficient management of resources calls for cost-effective solutions in reducing air pollution, greenhouse emissions and waste-water treatment. Global environmental issues could not be addressed without addressing the social dimensions. Major problems facing the world are: *climate change, bio-diversity loss, pollution of International waters, global warming, and ozone depletion. Today 1.7 billion people around the world do not have access to sanitation and potable water to drink and 3 billion people are breathing unsafe air.* All the countries of the world who were parties to the Rio Summit *must* initiate the process of achieving "environment, sustainable development objectives" agreed upon.

4. Global Climate Change

Human activities contribute to climate change and a bulk of it could be attributed to *fossil fuel* use particularly in developed countries.

The United Nations Framework Convention on climate change for developing countries is concerned with solving the problems of environment, resource management and poverty reduction. The reduction of Green House Gas (GHG) emissions through global partnership is the main issue. This could be achieved through reducing the emissions in such sectors as power, transportation, building and agriculture etc. Decrease in fuel consumption, better maintenance of vehicles and roads and traffic management would be helpful. Use of low energy substitutes, waste recycling could reduce carbon dioxide emissions. Reforestation, better cropping methods, use of bio-mass gasifiers, bio-methanation, solar-based rural electrification etc., are the other alternatives. Carbon sequestration could be increased through afforestation, greater bio-diversity etc.

5. Immediate Steps to Control Urban Pollution

Curbing vehicular and noise pollution, solid waste management, installation of Common Effluent Treatment Plants (CETP), flyash utilisation, levying pollution cess, use of low-sulphur diesel, synchronisation of traffic lights, diversion of traffic from highly polluted areas, curbing overloading of heavy vehicles, installation of fuel-gas desulphurisation plants for control of sulphur dioxide emissions, replacement of coal-based units with gas turbines on clean fuels-like Naphtha or natural gas would go a long way in combating city pollution.

6. Tightening Pollution Laws

Implementation of the following Environment and Pollution laws be scrupulously made :

1. The Air (Prevention and Control of Pollution) Act, 1981 as amended in 1987, and its Rules, 1982.
2. The Environment (Protection) Act, 1986; Rules, 1986.
3. Hazardous Wastes (Management and Handling) Rules, 1989.
4. Manufacture, Storage and Import of Hazardous Chemicals Rules, 1989.
5. The Water (Prevention and Control of Pollution) Act, 1974; Rules, 1975.
6. Water Cess Act, 1975, Rules 1978.
7. National Environment Tribunal Act, 1995.
8. Wildlife (Protection) Act, 1972.
9. Wildlife (Stock Declaration) Central Rules, 1973.
10. Wildlife (Protection) Licensing (Additional Matters), Rules, 1983.
11. Indian Forest Act, 1927; Forest (Conservation) Act, 1980; Rules 1981.

7. Recycling

Reusable Waste. Iron & Steel scrap, Glass bottles, Aluminium/brass scrap.

Recycling junk. Newspaper, plastic waste, Rubber tyres/tubes, Engine oil, lubricants, transformer oil.

Animal and Agricultural Waste. Use of anaerobic digestion, generating useful by-products like prawn food/algae, fertilizer pellets, recycled water for irrigation, conversion to electric power (from methane gas produced).

8. Cyclones, Floods & Disasters Mitigation

(a) Mitigation (preventive measures, reduction and protection, training and education)

(b) Preparedness (emergency management, direction and control, communications, shelters, warning systems, alerting and awareness)

(c) Response (Saving life, property, alleviation of suffering, medical, food aid etc.)

(d) Recovery (healing, restoration, reconstruction, rehabilitation).

Torrential rains, hurricane winds, storm surge (raising of sea waves) lead to cyclones, typhoons etc. Forecast is done with the help of Radars, weather satellites (INSAT - IB, INSAT - ID and polar orbiting satellites. When cyclone becomes apparent, the steps to be taken are : (1) Preparation of cyclone maps showing possible areas of submergence, water flow direction, dam burst possibilities, breach of embankments, breach dam locations for evacuation (2) Study of possible landfalls, storm surges (3) Dredging of rivers and estuaries (4) Construction of cyclone and storm proof houses (5) Long term soil and plant conservation, and coastal erosion measures.

Floods are caused by excess recipitation, aggradation of river beds, encroachment of flow planes, ice jams, blocking of stream courses, cyclones, deforestation. Forecasting of floods is carried out by Central Water Commission by monitoring river basins, hydrological and hydrometeorological data and issuing warnings. Flood management consists of structural measures like construction of dams, embankments, drainage channels etc. Non-structural measures include flood plane surveys, plane zoning, flood-proofing (bunds, raising levels), flood forecasting and flood insurance.

9. What is happening to the Earth — Heat Calamity

As the surface temperature of the Earth rises, the barrier increases the absorption factor of carbon dioxide modules, keeping the heat inside the atmosphere. This in turn leads to further heating so that the higher the temperature of the surface, the less it is cooled. In addition, intensive water evaporation increases the concentration of water vapours in the atmosphere, leading to the additional absorption of infrared radiation and heating. The concentration of carbon dioxide increases also due to its release from the ocean and minerals when the temperatures rise. This accelerates greenhouse effect. Ageing of the Bio-sphere, global climatic changes due to increased heat emission by *Global energetics* and carbon dioxide accumulation in the atmosphere promote heat catastrophic to the Earth. But the balancing factors are the oceans and accumulated humidity which counteract and save the Earth. *Earthquakes and Volcanoes* are thrown out due to the inner up-heavals that take place in the Earth. The inner core is a solid mass of extremely hot iron, 2400 km wide. Earth is made of four distinct layers : (1) A rocky crust, (2) A mantle of hot rock, (3) An outer core of liquid iron and (4) An inner core of solid iron. Studies on the Earth's interior by means of seismic waves and super computer revealed its geology, planetary formation and tidal forces produced by the satellite system on the "parent planet", and the breaking effect due to the planet's own majestic field in generating differential rotation.

Contamination of the atmosphere by dust from human activities, volcanoes and pollution can alter the *global heat balance* (by reflecting away incoming solar heat and diverting the escaping heat down towards the ground, making the surface warmer. Both Industry and Agriculture also add particles to the atmosphere leading to climate change.

U.N. Conventions have also stressed to control and ban hazardous wastes disposal, particularly the dumping of toxic wastes by the Industrial countries on the developing countries. There are other UN documents on the transboundary effects on industrial accidents and the Global Environment Facility.

The recent NAM's declaration at New Delhi has also taken stock of the progress of implementation of the Agenda 21 of the Rio Summit for which the Industrialised/Developed countries of the West have yet to fulfil their role/commitments and discharge their part of the obligation in respect of pollution control and the associated hazards, at least by the time when the next Earth Summit meets.

INDEX